MW00478783

BROWN ACRES

BROWN

AN INTIMATE HISTORY
OF THE

The original sewage screening plant was established at Hyperion. 1925

CRES

LOS ANGELES SEWERS

Anna Sklar

Brown Acres
An Intimate History of the Los Angeles Sewers
Copyright © 2008 by Anna Sklar

Designed by Amy Inouye, www.futurestudio.com

10 9 8 7 6 5 4 3 2 1

ISBN-13 978-1-883318-75-8

Use of lyric from "Song of the Sewer" courtesy of Songsmiths and Jackie Gleason Enterprises, LLC.

LIBRARY OF CONGRESS CATALOGING-IN-PUBLICATION DATA

Sklar, Anna, 1940-
 Brown acres : an intimate history of the Los Angeles sewers / by Anna Sklar.
  p. cm.
 Summary: "With more than fifty photographs, diagrams and maps, Brown Acres is the first historical narrative to detail any world-class city's sewer system—complete with the relationship between headstrong politicians and the reformers seeking to "heal the bay" after a century of pollution and contamination. Brown Acres provides a unique look at the underground history of Los Angeles" —Provided by publisher.
 Includes bibliographical references and index.
 ISBN 978-1-883318-75-8 (trade paper : alk. paper)
  1. Sewerage–California–Los Angeles Metropolitain Area–History. 2. Water–Pollution–California–Los Angeles Metropolitain Area–History. 3. Environmental protection–California–Los Angeles Metropolitain Area–History. I. Title.

TD525.L7S55 2008
363.72'8493–dc22

                         2008006189

Printed in the United States of America

ANGEL CITY PRESS
2118 Wilshire Blvd. #880
Santa Monica, California 90403
310.395.9982
ANGEL CITY PRESS   www.angelcitypress.com

For my children, Jeffrey and Tracy

The author gratefully acknowledges the support of the
Los Angeles City Historical Society, under whose auspices this book was
written, and of the John Randolph Haynes and Dora Haynes Foundation
for a grant to complete the project.

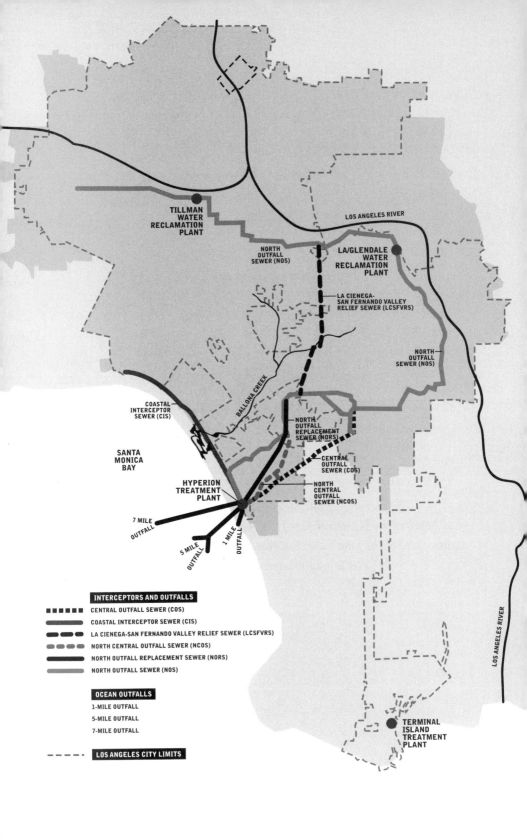

TILLMAN
WATER
RECLAMATION
PLANT

LOS ANGELES RIVER

NORTH
OUTFALL
SEWER (NOS)

LA/GLENDALE
WATER
RECLAMATION
PLANT

LA CIENEGA-
SAN FERNANDO VALLEY
RELIEF SEWER (LCSFVRS)

NORTH
OUTFALL
SEWER (NOS)

COASTAL
INTERCEPTOR
SEWER (CIS)

BALLONA CREEK

NORTH
OUTFALL
REPLACEMENT
SEWER (NORS)

SANTA
MONICA
BAY

CENTRAL
OUTFALL
SEWER (COS)

HYPERION
TREATMENT
PLANT

NORTH
CENTRAL
OUTFALL
SEWER (NCOS)

7 MILE
OUTFALL

5 MILE
OUTFALL

1 MILE
OUTFALL

LOS ANGELES RIVER

INTERCEPTORS AND OUTFALLS

██████ CENTRAL OUTFALL SEWER (COS)
━━━━━━ COASTAL INTERCEPTOR SEWER (CIS)
━━━━━━ LA CIENEGA-SAN FERNANDO VALLEY RELIEF SEWER (LCSFVRS)
━━━━━━ NORTH CENTRAL OUTFALL SEWER (NCOS)
━━━━━━ NORTH OUTFALL REPLACEMENT SEWER (NORS)
━━━━━━ NORTH OUTFALL SEWER (NOS)

OCEAN OUTFALLS

1-MILE OUTFALL
5-MILE OUTFALL
7-MILE OUTFALL

- - - - - **LOS ANGELES CITY LIMITS**

TERMINAL
ISLAND
TREATMENT
PLANT

CONTENTS

Introduction 9

Chapter One: Rivers of Darkness 12
Chapter Two: Sewers for the Pueblo 18
Chapter Three: A Sewer System Is Born 25
Chapter Four: Sewer to the Sea 36
Chapter Five: Sewage Woes 44
Chapter Six: More Sewage Misery 55
Chapter Seven: Dilution Is the Solution to Pollution 66
GALLERY   Early Sewerage: 1887–1945 79
Chapter Eight: Pollution of the Santa Monica Bay 93
Chapter Nine: A Modern Treatment Plant 105
Chapter Ten: An Interim Solution to Pollution 121
Chapter Eleven: The Clean Water Battle 136
GALLERY   Decades of Progress: 1946–2008 151
Chapter Twelve: Do-Gooders Get Organized 178
Chapter Thirteen: Healing the Bay 182
Chapter Fourteen: No More Brown Acres 198

Afterword 209
Acknowledgments 213
Bibliography 217
Index 227
Photo Credits 232

Introduction

We sing the song of the sewer
Of the sewer we sing this song.
Together we stand
With shovel in hand
To keep things floating along.
—"Song of the Sewer," created for *The Honeymooners*

I first heard about the "Song of the Sewer" a few years ago, when I began working on this book. Best friend of overweight, irascible bus driver Ralph Kramden, the put-upon sewer worker Ed Norton would sing the song and we'd snicker. Norton worked in the sewers, but he never seemed to carry evidence of his work home. That's probably because the last thing any of us wants to think about is what lies in the pipes beneath our feet.

◁ Sewage poured directly into Santa Monica Bay. circa 1935

I learned more than I thought I would ever want to know about sewage in 1983 when I went to work for the Los Angeles Public Works Department, after working for the city's Cultural Affairs Department. Soon, I found myself writing and talking about the city sewers. Before then, as a reporter, I had covered hundreds of news stories but had never even talked to a sewer engineer, let alone explored the vast network of wastewater pipes and sewage treatment plants that handle the waste of more than four million people who live in L.A.

Soon I was swept up in a swirl of controversy as angry environmentalists accused city officials of deliberately polluting Santa Monica Bay, which they insisted the city had been doing for decades. The beleaguered city politicians and engineers were just as convinced that they were properly treating the city's wastewater. Eventually, the politicians and the engineers joined the environmentalists and successfully convinced voters to approve more than three billion dollars to fix the ailing sewer system.

In 1994, shortly before I left the city payroll, I discovered that one hundred years earlier engineers had built the first sewer pipe to a sand-dune-filled beach, twelve and one-half miles from the city limits. I was intrigued and soon began doing some research, in a more or less desultory fashion, thinking that I might one day write a book about the history of the city's sewer system. I never really lost interest, but no one else seemed particularly interested in the subject, so I went on to other things.

Two years ago, I was chatting with Hynda Rudd, then president of the Los Angeles City Historical Society, formerly the city archivist. She insisted that I write the book, and with her support and that of the historical society, I received a grant from the John Randolph Haynes and Dora Haynes Foundation, which helped me complete my research and write this book.

We all think, "ecch," or "yuck," when someone mentions sewage. We have no reluctance, however, to use words such as "sh*t" and "crap" when we drop something or injure ourselves. We express amazement with "no sh*t." These words are ubiquitous in the English language. Although they are not acceptable in "polite society," few can avoid the inadvertent expressive expletive when out-of-the-ordinary things happen. Yet we rarely use words such as "excrement," "feces," or even "ordure." We prefer, like Oprah, to say "poop" when talking about that yucky stuff in the sewers. It's as if, as one cynical engineer said, "they think the sewage fairy takes it away."

Still, since you picked up this book, you, the reader, are probably just a bit curious about how our sewer system works and what went on before Los Angeles became a leader in "exceptional" (as government officials like to say) treatment of its wastewater, also known as sewage.

Before turning to a somewhat contentious and tumultuous sewage history going back to 1860, the reader will want to know more about how and why it became such a problem in the late twentieth century. So, we begin with a brief history of sewerage as it developed through the ages.

—Anna Sklar
2008

CHAPTER ONE

Rivers of Darkness

ivers of darkness flow beneath the city streets of Los Angeles. Closed and hidden, the sewers are protectors against the silent armies of bacteria and viruses that would otherwise lay waste to the population. One hundred fifty years ago residents were using open ditches to dispose of their body waste. Yet, five thousand years ago in Mesopotamia, the Babylonians created a sewer system that captured storm water and human waste that flushed to drains made of sun-baked bricks or cut stone. Many homes contained a small room with an opening in the floor over which people sat to relieve themselves. The wastes fell through the opening into a perforated, circular cesspool located under the house. These early cesspools were made of baked perforated clay and were stacked atop one another. They were often filled with pieces of broken pottery to allow for better percolation.

◁ Sewer workers repair the Central Outfall Sewer in 1948.

In the same period, in Pakistan, drainage systems were located in the streets. As Jon C. Schladweiler, historian of the Arizona Water and Pollution Control Association, wrote in "Tracking Down the Roots of Our Sanitary Sewers" on the sewerhistory.org Web site, "Liquids entered brick-lined cesspools or were carried to the local river for discharge." Homes had bathrooms and latrines built next to each other on the street side of the home. Water was used for flushing latrines. Around 3000 B.C., the Aegean civilization built drainage systems of terra-cotta pipe that conveyed storm water and human wastes. Latrines were flushed with water from large jars. In ancient Greece, storm water and human waste was carried to basins outside of towns. And, as Schladweiler explained, it "was then conveyed through brick-lined conduits to fields to irrigate and fertilize fruit orchards and field crops.

The Romans, however, built the best-known of ancient sewerage systems. The complex aqueduct system they built in 510 B.C. included the eleven-foot-by-twelve-foot *Cloaca Maxima* (the main drain) that drained to the Tiber River. Sewage facilities included "rooms of easement" where public officials would sit on latrines and discuss business. These latrines were constructed of elongated rectangular platforms with seats. Although human wastes were often thrown into the streets, regular street washing caused the wastes to end up in the drain system.

Besides public latrines and the few houses connected to the drains, huge vases were provided for use at the edge of towns at entrance roads, while vendors worked the streets providing access to pottery jars and "modesty capes." Cloacina was the patron goddess of the *Cloaca Maxima* and the city's overall sewer system. Over time, the Romans came to think of her as the goddess of purity as well as the goddess of filth and as the protector of sexual intercourse.

When the Roman Empire fell, the barbarians—to whom body cleanliness was unnecessary—rushed in and destroyed most of the sewerage works. With the dawning of the Middle Ages in the fifth century a new age of romance was ushered in: uncleanliness was next to godliness. Moats surrounding castles were more than defense arrangements; they often were filled with human waste. Enemy combatants might drown, but they probably would have died anyway from the massive amount of contamination in the water. The vast majority emptied their waste from buckets into the roads and streets. Filthy water coursed into streams and rivers that became clogged and polluted. For the few who could afford housing and were without castles and moats, cesspools proliferated. Whatever cesspools existed were rarely cleaned and frequently overflowed into the public streets.

By the fifteenth century these older cesspools often leaked into adjacent streets, where street sweepers removed the solid waste. They and cesspool cleaners brought their refuse to dumps located throughout the city. In subsequent centuries, privy vaults (sealed underground containers) joined the cesspools as the primary recipients for body waste. Sewers were built in the eighteenth century, but they emptied into rivers that supplied the drinking water for people living nearby.

As population swelled in urban centers such as Paris and London, by the 1840s and 1850s the polluted water had migrated to wells from which everybody got their drinking water. Soon, cholera and typhoid epidemics swept through the urban centers of England and France, killing thousands. Although at first many doctors were convinced that epidemics were caused by "miasma" in the air, a determined and heroic London physician, Dr. John Snow, convinced the authorities that bacteria in the putrid waste carried by sewers were causing the plagues. Nevertheless, it took the "Great Stink" of London in 1858 to finally force city officials to build epic grand sewers to bypass nearby rivers and empty the sewage into the Thames downstream and east of London. French engineer Eugene Belgrand soon developed a similar, even more massive sewer system for Paris that was completed in 1878. The cholera epidemics ceased.

The science of sewers contains its own vocabulary. Sewerage refers to the system of sewers, but just as often engineers will interchangeably use the term sewers as a reference.

As early as 1887, one of the first city engineers recognized that rainstorms in Southern California were unpredictable and could last days, even weeks. So, to avoid overloading, from the outset of the Los Angeles sewer system the sewers were designed to be closed conduits to carry wastewater from households and businesses to the ocean for disposal.

In Los Angeles, storm drains are not connected to the city's sewer system. Catch basins in city streets collect water that flows into gutters and conveys it to huge concrete storm drains. The storm drains sometimes empty directly to the ocean, but more often convey street runoff to streams such as Ballona Creek and the Los Angeles River, which then empty into Santa Monica Bay.

Drains from homes and businesses are connected to city street sewers by small pipes about four inches in diameter, called service laterals or lateral pipelines. These small-diameter sewers connect to city sewer mains that are six to eight inches in diameter. The mains connect to laterals that are usually sixteen inches in diameter. The further into the network, the larger the pipes, as evoked by the terms used to describe them: mains, collectors, trunk lines, interceptors. (When it comes to sewage, terms are sometimes relative: "outfall" is downstream from some point of reference and "collector" is upstream.) Interceptors, the largest outfalls, carry the sewage to four L.A. treatment plants: the Hyperion Treatment Plant, the Terminal Island Treatment Plant, the Donald C. Tillman Water Reclamation Plant and the Los Angeles-Glendale Water Reclamation Plant.

All sewers are constructed on grade so that whenever possible the sewage flow can be carried by gravity to treatment plants. Major outfalls range from as little as two feet in diameter upstream to twelve and one-half feet wide downstream. Where necessary, pumping stations lift the sewage to a higher elevation to allow it to continue to flow. Some sewers are laid as little as two feet below ground, while a few are placed almost 125 feet deep to avoid utility lines and other underground structures. To ensure the continued flow by gravity when the grade runs through a hill, the sewer line may be laid deep in tunnels.

The L.A. storm drains that are mostly open-ended concrete conduits often provide a backdrop for motion pictures, but the closed sewers of Los Angeles remain a mystery. However, the treatment plants, with their massive sets of pipes and unusual physical attributes, have often served as locations for various science-fiction movies, television programs and the occasional commercial. But they are never represented as sewage treatment plants. Sewers and the treatment plants represent the loss of our bodily fluids and solids. As Cecil Lue-Hing, an expert on municipal waste, put it, "It's ingrained in us that our body waste is bad stuff. Don't show it to me, once it's left my body."

On the other hand, water, pure water, represents power, as has been demonstrated in the dozens of books indelibly linking the two images in our minds. Yet, the pipes that convey the city's excreta and other effluent, as well as the plants that treat the waste, have their own intrigue and play a vital role in the infrastructure of the city. And engineers design and build this infrastructure—without which we would be living in the Middle Ages.

Like the freeways above the city streets that connect people, homes and industry to the cities in which they exist, the sewers below the streets serve the same purpose. With few exceptions they follow the natural course of gravity through the Los Angeles Basin as they collect the wastewater from far-flung communities. From Wilmington and San Pedro, the west San Fernando Valley through Burbank, to Glendale, East Los Angeles, Boyle Heights, Beverly Hills, West Hollywood, Culver City and El Segundo, 6700 miles of sewers convey 450 million gallons of wastewater per day from more than four million people to four treatment plants.

Like the city itself, the sewer system has grown in spurts, mirroring housing booms from the late 1880s to the present. Political wars, recalcitrant voters, visionary and charismatic (often contentious) engineers and corrupt politicians in turn took the stage as residents of the small pueblo by the Los Angeles River took up the challenge of sewerage.

CHAPTER TWO

Sewers for the Pueblo

When Los Angeles was incorporated in 1850, shortly before California was admitted to the Union, there were no sewers in the pueblo of twenty-eight square miles and little more than sixteen hundred residents. *Zanjas*, open ditches perpendicular to the Los Angeles River, served as both water supply for the town and disposal sites for garbage and other waste, including the contents of cesspools and privy vaults.

The first sewer in L.A., according to contemporary historian Harris Newmark, was built sometime before 1861. He described it as "a small square wooden pipe" that lay underneath Commercial Street near the main plaza. After thirty days of continuous rain that deluged and flooded the town, dozens of homes were destroyed, so it's not surprising that "in February or March," as Newmark reported, "the sewer apparently burst and flooded the common

◄ Downtown Los Angeles in the1860s.

yard adjacent to the Bella Union Hotel." According to newcomer Alice Eaton, the hotel "was a watering hole for many men of the pueblo," which she said "was wild in those days; saloons and drunkards everywhere. Killings almost every day, too, and lynching." By all accounts it was a disreputable frontier town—a heterogenous mix of native Californians, African American settlers, recent immigrants from Mexico, Native Americans, a few Chinese and hundreds of adventurous Americans seeking their fortune in the small town by the Los Angeles River.

A sewage system was not high on a list of necessities for the residents. At least not right away. But shortly after the first private sewer exploded, its replacement was built with city funds. "This great sewer," later recalled local raconteur Horace Bell in *Further Reminiscences of a Ranger*, "which cost more than five thousand dollars, was three feet wide, one foot deep and laid one foot underground, dumped into a cesspool, and only one man's property was connected with it." Bell's disgust with the sewer was soon matched by city officials. The local court ordered it closed, while the cesspool to which it was connected was filled with earth.

The first three appointed city engineers had little to do with the development of sewers. Their primary obligations were to map the town, construct streets and sidewalks, and install street lights. After the Civil War, more people settled in the town and within a few years, it became clear to local officials that along with other amenities, a sewer system was essential. The Common Council (the form of government before a charter change in 1889 created the current system of a city council) appointed the first sewer committee in 1869. The committee's first assignment was to devise a sewer outlet for the new Commercial Street sewer, built to replace the first sewer. Following its successful construction, Robert L. Lecouvreur was appointed city engineer and surveyor in 1869.

Although it does not appear that he did anything to connect the new sewers being built in the town, shortly before leaving office in 1870 Lecouvreur prepared a seventeen-page model for an ideal sewage system for the more than 5,700 residents of the expanding pueblo. Noting the unique terrain of the city and recent published reports by engineers on the East Coast, he advocated that sewers be smaller than those in Europe. He wrote that if the sewer were constructed proportionately for depth and velocity, "its ability to keep itself clean will be confined to high rates of inclination, it is important therefore to have [the sewer be] as small as the service to be required of it, will admit."

Lecouvreur was the first to advocate separating the storm drains from the sewers, suggesting that any "attempt to fit the size of the sewers in this city to an accommodation of those extraordinary storm waters which happily occur only at exceedingly long intervals, would be little short of idiotic, nor exists there the remotest necessity for doing so." (Subsequent engineers agreed with him.) He thought, however, that it would be a good idea to dump the sewage into the Los Angeles River south of the pueblo. He was not the first or the last to suggest this course for the city's sewage. Fortunately the council did not act on that recommendation. However, the residents were not prepared to tax themselves for a sewer system. Lecouvreur's lengthy report languished for over a century in the city archives, ignored and neglected by city officials.

The council later appointed several sewer committees: some were formed to study the extension of the sewer system, others for sewage disposal and some to find a sewerage route south of Washington Boulevard. These men issued new regulations and established which materials should be used for construction of sewers: brick, wood and, where possible, vitrified clay; they were to be egg-shaped in cross-section, two feet wide with eight-inch-thick walls. Property owners connected to the sewers were assessed for the costs of construction.

Despite new sewer ordinances, sewerage proceeded in a haphazard fashion. In April of 1874, the frustrated council members ordered an end to the "noxious nuisance" of the wooden sewer on Commercial Street. A special committee was appointed to determine the financial damage to property owners caused by this and other city sewers. Claims for damages to property fronting the sewers were paid from the city treasury. Property owners under whose land sewer lines were constructed objected strenuously to the city's right of way. This was a time of rough development, when business enterprise was more important than sanitation.

In 1876, the Southern Pacific Railway established a line to Los Angeles County, bringing in hundreds of new settlers. A new main sewer was built along Los Angeles Street to meet the needs of the growing city. It deposited the waste in a cesspool in the southwest part of the city, where open land was used for vegetable farming.

As population increased, the council now required that people's privies, vaults and cesspools be connected to a city sewer, if one existed. Where no city sewer was available, residents were required to build "vaults, at least eight feet deep, with an escape pipe extending at least two and one-half feet above the top of the closet." To prevent sewage gas from spreading over the town, the council required securely covered traps to be installed for all cesspools or sinks receiving drainage from kitchens. *Zanjas* were becoming clogged with runoff from drains and cesspools, so new ordinances were approved forbidding discharge from them into the streets. Washing and bathing in the *zanjas* was also forbidden. Nevertheless, as a water system had not yet been developed, many people continued to bathe and to wash their dishes and clothes in the *zanjas*.

During this period, sanitation throughout the United States was in a general state of underdevelopment. Although little is known about sewerage in America before the 1870s, most sewers were constructed by individuals at little expense and without any public supervision.

Early sewers built to replace privy vaults and cesspools were often built on a level grade and would become congested. The sludge that accumulated as the sewage decomposed caused a disgusting widespread stench. Because of the failure of cesspools that often overflowed their borders, disease was commonplace in the opening decades of the nineteenth century, especially in the developing urban areas. By mid-century, American engineers began to emulate the English and French engineers who had discovered the necessity of constructing sewers with a sufficient grade to carry the effluent away from urban centers into rivers downstream from water supplies.

As pioneering American engineers Leonard Metcalf and Harrison P. Eddy wrote, before the cholera and yellow fever epidemics of the mid-1850s, "sewerage was designed to carry off only surface waters, not human waste which was for the most part buried in cesspits or cesspools." By the late 1870s and 1880s, with cholera and typhus plaguing dense new urban centers, sanitary engineers in cities such as Atlanta, Chicago, Philadelphia and Memphis emulated European engineers and began rerouting sewers away from rivers and streams.

Although a few wooden sewers were still in operation in Los Angeles by the 1880s, all new sewers were built of brick, cement pipe or burnt glazed tile. The new sewers were no longer one size from beginning to end. Most were wider at the downstream end to accommodate larger flows there. In this decade the city doubled its population from 5,700 to eleven thousand.

To avoid stagnant open pools of raw sewage, in 1883 the council granted to the South Side Irrigation Company free rights for eighteen years to all the sewage emptying into the cesspool in the southwest corner of the city. The company then sold the sewage as fertilizer to Nadeau Vineyards and other large farms.

By 1884, the three main sewers under Alameda Street, Los Angeles Street and San Pedro Street were decaying. Soon, residents began complaining to the sewer committee about the terrible stink. However, dry and thinly populated Los Angeles escaped the common plagues of cholera and typhus that decimated congested urban cities in the Northeast United States. Neither the city's health officer nor its residents pressed for a system to remove the sewage from the city.

George Knox, who was city surveyor and city engineer from 1884 to 1886, did little to alleviate the sewage problems. The council's sewer committee served as the enforcement arm for the city but its only remedy was to take individual business owners to court.

City officials had begun to look for a solution to the ever-widening sewage problem when the Santa Fe Railway completed a spur to Los Angeles in March 1886. Soon, hundreds of thousands of tourists arrived in town via the new railway, many of whom settled permanently in the city. The Southern Pacific and the Santa Fe soon began competing for fares; as each lowered its rate, the other rushed to follow suit. The railroad rate war exploded in fury in March 1887. By the afternoon of March 6, fares reached a low of one dollar for travelers from the Midwest and the Mississippi Valley. Although fares rose somewhat in the following months, they remained below twenty-five dollars for the rest of the year.

By the middle of March 1887, the land boom that at one point attracted thousands of visitors daily reached hysterical land-buying proportions never seen before. Land changed hands rapidly, often turning over too quickly for the deeds to be registered at the land office. More than seventy thousand new residents settled in Los Angeles in 1886 and 1887. One hundred new homes a month were built in 1886. Newcomers approved public works projects to pave the streets, build the suburban railway lines and provide electric lighting for the streets. In the short period from August to December of 1887 the post office handled mail for more than two hundred thousand transients. Before the boom, there were no paved streets; by the end of 1890 there were eighty-seven miles of paved streets and eighty miles of interurban railway track. The dusty pueblo of multiple ethnic groups was transformed into a predominately white city, populated by middle-class people who changed its tone.

In December 1885, Fred Eaton was elected city engineer. After taking office in January, he began working on a new sewer system for the city. While laying down street sewers in March 1886, the council requested a design for a combined sewage outfall. Eaton responded, like Lecouvreur before him, that the city should not build a combined storm and sewage system because the region's unpredictable rain could overwhelm such a system. By early 1887 he had shaped legislation to improve the sewers, especially by eliminating rainwater connections that overloaded the sewers when more than twenty-two inches of rain fell in that year's wet winter.

In March 1887, shortly after the railway rate war reached its fever pitch, Mayor William Workman asked Eaton to prepare recommendations for a sewer system for all of Los Angeles. Eaton, the first city engineer with an extensive background in water management, delivered his plan within weeks of the request.

CHAPTER THREE

A Sewer System Is Born

Native-born Angeleno Fred Eaton designed the first comprehensive sewer system for the city in 1887, but it would be seven years before his plan would be fully implemented. He started his engineering career at fifteen, when he went to work as an apprentice for the privately owned Los Angeles City Water Company. Five years later he became the company's general superintendent. While superintendent, he befriended William Mulholland, who came to work for the company in 1878 and later built the Los Angeles Aqueduct. Both men studied engineering while working for the water company and maintained a close friendship for the next thirty years.

Eaton was a popular member of the Republican Party and a respected engineer who had twice been elected city engineer, yet the council decided to ask an outside expert to consult with Eaton before he completed a sewage plan. Colonel George E. Waring was the first in a succession of consultants city officials turned to in order to solve such issues.

The city approached Waring, a well-known sanitarian, self-taught engineer, lecturer and rather flamboyant self- promoter who had earlier designed systems for Memphis and Brooklyn. Horrific yellow fever outbreaks in Memphis had killed thousands of people in the 1870s. These epidemics ended after installation of Waring's sewer design, and he became famous for of his work. In 1886 he completed a two-volume report on the nation's sanitation systems for the United States Bureau of the Census. Waring appeared to be the perfect person to consult about Eaton's plans. So, on April 26, 1887, Eaton suggested that the city pay Waring five hundred dollars to work with him on designing the system.

On May 3, Eaton offered his now slightly modified plan, with which Waring agreed. The plan called for an interior sewerage system and an outfall to the sea. The interior system included three major interceptors that would cover the entire city of twenty-nine square miles (an additional square mile south of the city had been annexed in 1859). Laterals would be built to ensure that sewage from East Los Angeles, Boyle Heights, the northern, southern and western parts of the city would all be connected to the new system. The sewage would then be discharged to a settling tank at Agricultural Park, a somewhat disreputable agricultural fairground of 160 acres that included dozens of saloons and was used for horse races, brothels and other sporting attractions. The park was in the area that today encompasses Exposition Park and the Los Angeles Coliseum.

The outfall leaving the settling tank at Agricultural Park would extend twelve miles to the ocean, discharging into the surf south of Ballona Creek at what is today Dockweiler Beach. Eaton suggested that the sewer be designed to provide sewage water to farmers during the summer months to irrigate land under which the outfall would be built. During the winter, the wastewater could be delivered directly to the ocean.

On May 12, Mayor Workman and some council members joined Eaton for a survey of the land under which he proposed the sewer should be laid. The property was owned by Hugh Vail and Dan McFarland who, along with other businessmen, had recently purchased several thousand acres of the Centinela Ranch. The men planned to subdivide most of the land in order to create the city of Inglewood. The rest would be left for farming. According to published reports of the trip, the owners of the land appeared eager to have the outfall on their property. If the outfall allowed sewage water to be carried to their farmland for fertilizer and summer irrigation, farming costs would be reduced. On May 14, Colonel Waring announced his support of Eaton's proposal, which both men believed could serve a population of two hundred thousand. Meanwhile, it became clear that because of the potential cost of the outfall several elected officials and leading businessmen were unhappy with Eaton's plan.

A second sewer engineering expert, Dr. Rudolf Hering, contacted the council in June and offered, for twelve hundred dollars, to look at Eaton's system design. Hering was a trained engineer and had worked from 1876 to 1880 as assistant city engineer in Philadelphia. Like Waring, he was the author of several articles about sewerage problems. In 1877, in the midst of a typhoid epidemic in Chicago, he had designed a sewer system for that city, which at the time deposited its sewage into Lake Michigan. He built a sewage canal, diluted the sewage with water from the lake and diverted the sewage to the Des Plaines River, which flowed into the Illinois, a tributary of the Mississippi. At the time, dilution was the leading solution to sewage pollution.

In many respects Hering was even more celebrated and well-regarded among sanitarians than Waring. He had recently returned from a government-sponsored tour of Europe to evaluate that continent's sewage systems. Hering reviewed and slightly amended Eaton's plan for new interceptor sewers within the city borders. Together, they presented a map of the sewer system for the city, as well as several additional maps that traced the course of the proposed twelve-mile outfall.

Hering's typed, single-spaced, twenty-three-page report offered the elected officials an opportunity to evaluate all the possible methods of solving their sewage problems. He assumed their knowledge to be minimal and spelled out their options carefully. One option, he wrote, would be to divert the sewage into the ocean or a large water course. Secondly, he said, it could be spread over large areas of land and, by ordinary methods of irrigation, utilized for agricultural purposes, or it could be confined to smaller areas of land and purified by filtration through the soil. Finally, he said, it could be treated by a chemical process which would cause the suspended and some of the dissolved putrescible matter to be precipitated, leaving the liquid sufficiently pure to be turned into the natural water courses. He dismissed this chemical method as too costly and less efficient than the first two, both of which were included in Eaton's plan.

Most important, Hering agreed that the outfall from the city limits should follow a natural gravity flow to the ocean, as Eaton wanted, about two and one-half miles south of Ballona Harbor. He also recommended that the sewage should be strained and released only at high tide. In England and in U.S. Eastern Seaboard cities, Hering wrote, "foul waters are taken out by the receding tide and so thoroughly dispersed and diluted, that they soon become unnoticeable."

Hering advocated a discharge point several hundred feet from shore, adding that the sewage should be stored in reservoirs and allowed to flow out only during the first two hours of outgoing tide. Eaton did not include Hering's idea about storing or screening the sewage in the final plan he submitted to the council. Both Hering and Eaton, however, agreed on the need for separate storm drain and sewer systems. They recognized that the periodic drenching downpours of rain in the city could cause combined sewers to burst from the excess pressure.

In minute detail, Hering documented the proposed size of the sewer, placement of manholes and ventilation shafts. He wrote that "street sewers should be no less than eight inches," and suggested that the grade for the pipes "should be at least six inches per hundred feet, which is ample to secure a cleansing velocity of the water. Main or intercepting sewers should be twelve inches," he concluded. Both Hering and Eaton wanted the outfall to "be composed of two interlocking rings of bricks set in mortar inside the clay pipe." The precise nature of his comments was meant to reassure non-technicians with his detailed understanding of sewer construction.

On September 12, 1887, Fred Eaton presented his amended sewer plan. The total cost for the internal sewer system, the storm drains and the outfall sewer would be a million dollars. In describing the outfall, he wrote that it "should be about 11.5 miles long; egg-shaped rather than circular to increase efficiency." The outfall, four feet by five feet, six inches, would be considerably larger than any of the interceptors, which he wrote "would allow for over two hundred thousand persons." Eaton was no fan of wooden pipes, so he listed all the material to be used: extra-common invert bricks, common hard-burnt brick, vitrified pipe of three different lengths, eighty-one tons of cast iron pipe, thousands of cubic yards of concrete, eighteen thousand barrels of Portland cement and six thousand cubic yards of mortar sand. He specified the number of manholes and flush gates to be installed, and laid out the exact number of feet to be built in tunnel through dry sand.

Unfortunately, Eaton's careful assessment of the city's needs met with instant opposition. Mayor Workman vetoed the plan in October, after which a fierce public debate about the plan raged. Voters had no quarrel with Eaton's plans for the interior system and eventually approved funds to build these sewers in March 1890. However, the debate about the outfall—whether it was needed at all, where it should be placed and whether the city should be responsible for handling the sewage after it was brought by the interceptors to the southern part of the city—remained unresolved for five years. The *Los Angeles Times* printed a hundred letters to the editor—more than on any other topic during the 1880s—about how to dispose of the sewage.

Many businessmen, property owners, newspaper editors and council members argued that at four hundred thousand dollars the sewer outfall was too expensive. In their minds, spending several hundred thousand dollars to throw away the sewage into the ocean was essentially a waste of money. At the time Los Angeles had the most extensive sewage farming of any city in the United States; the sewage of about seventeen thousand people was then being applied to more than sixteen hundred acres of land.

In December 1887, Eaton left office to pursue real estate and other business interests. William Lambie then served one year and offered a slightly different option for the outfall sewer. His plan met with little approval and the debate went on. Meanwhile, travel writers hired by the railroads to promote tourism to Southern California ignored the growing concerns about sewage. Ironically, however, in a popular guidebook to Southern California, published in 1888 in the midst of the sewage debate, Drs. Walter Lindley and Joseph Widney cast a rosy glow on the city's sewage disposal. "The sewage," they wrote, "is taken to a sewage farm where it is conducted from the outlet of the main sewer by ditches—one day to one orchard, vegetable-garden or vineyard, and the next day to another. As fast as it is received upon the land," they wrote, "it is plowed under and thus covered with earth, the best-known disinfectant."

In June 1888 South Side Irrigation agreed to build a twenty-two-inch cement pipe from the city's main sewer ending at San Pedro and Jefferson that extended several miles south of the Vernon district. They then sold the sewage to farmers south of the city.

Four days after the appearance of South Side Irrigation before the council, a reporter for the *Herald* visited Eaton, who was briefly back in the city, and asked him what he thought of a new sewage plan which would pipe sewage through a thirty-inch cement pipe to Ballona Slough—into which sewage would be dumped and then be allowed to drift along into the ocean. Eaton was annoyed that the city was still discussing alternate routes for the sewage. He dismissed the proposal because he felt that "a thirty-inch cement pipe is not adapted to conveyance of sewage; in time it cracks and clogs would occur." He reminded the reporter that his own plan called for enough capacity for a city of two hundred thousand. The thirty-inch pipe, he insisted, "would be adequate for a city of no more than fifty thousand inhabitants."

The patrician, rather stiff-necked Eaton reminded the reporter of his earlier efforts with Vail and McFarland, and said he had been asked by Mayor Workman to approach them again to learn of their possible interest. Eaton said that earlier the partners asked for a perpetual grant of the sewage, and for its use they offered the city a bonus of one hundred thousand dollars or 3,300 acres of land, beyond which they would deed a right of way over their property. Now, he said, "They declined and McFarland told me that about one-half of the land he had offered had been sold for two hundred fifty thousand dollars, and he expected to receive as much for the remainder. So you see that their offer, if accepted, would have paid for the entire sewer system of the city. The only alternative that appears to be left now is the Ballona route, but this will entail an entire waste of the sewage, as there are no lands down there that are suitable to receive it, inasmuch as there must be a sub-drainage or else its good effects are lost."

In February 1889, with a new charter making the post of city engineer an elected position, Eaton ran unopposed as a Republican and was elected by acclamation. He again took up the fight to site the outfall. He now acquiesced to pressure for a cheaper route and approved of the Ballona route to Ballona Creek for the proposed outfall sewer. The new route would go under land owned by prosperous businessman Daniel Freeman. Freeman, however, was adamantly opposed to any sewage lines being laid across his property.

Freeman had originally leased his property, the Centinela Ranch, in 1873 after coming to California from Canada. After the drought of 1875 killed twenty-two thousand head of his sheep, he planted 640 acres of barley. The plantings were extended and by 1880 the Centinela was producing a million bushels a year, some of which was shipped to New York and London. By the time he sold more than half his acreage to Vail and McFarland, Freeman was one of the wealthiest businessmen in Los Angeles. He was also an influential member of the recently formed chamber of commerce

In May 1889, Eaton and members of the newly established Los Angeles City Council met with the East Los Angeles Chamber of Commerce in an effort to develop support for the new system. At the conclusion of the meeting the chamber expressed complete confidence in his efforts. Others were less enthusiastic.

New opposition developed from the Santa Monica Board of Trade and the Ballona Harbor Company. The Santa Monica Board was beginning a battle to convince Congress to appropriate money to build a deep-water harbor there rather than in San Pedro, and did not want a sewage outlet, even a few miles south, that could potentially damage its bid. The Ballona Harbor Improvement Company was in the midst of a major development program for recreational activities near the slough and was equally dismayed by the city's new plan. Abbott Kinney of Santa Monica, who would later build his famous Venice canals, opposed the Ballona route and the *Los Angeles Herald* argued that the city of Santa Monica would have cause for damages if the sewer were built to the ocean.

As sewage scented the air, righteous citizens debated the merits of various methods of its disposal. These debates were fully covered in the pages of both the *Los Angeles Times* and the *Los Angeles Herald*, two of the largest newspapers at the time. Former Mayor Edward Spence was convinced that the beaches would be contaminated by sewage from the outfall. To prove his point, he noted that the beaches in Scotland and at Dublin, Ireland were polluted. At the Thames near London, he argued, the waters were rendered impure for great distances and he believed that there was no question that the sewage pouring into the ocean would destroy the beach for dozens or scores of miles on either side of the point of discharge. Although his concerns were not seriously considered at the time, fifty years later the California Department of Health Services would quarantine the beaches from Venice to Manhattan Beach for almost ten years because of untreated ocean sewage pollution.

The arguments increased and varied. One man claimed that the outfall sewer was a consummate fraud. Many people believed that building a sewer following the route of the Los Angeles River would be better and cheaper, with an outfall to the ocean at Long Beach. Then there were those like Judge Stephens who believed in simplicity. "We have no need for an outfall sewer for years to come. We can just pay Nadeau's Vineyard and Dr. J. Hannon money and they'll take it all."

In 1889 proposals were offered from several companies, including one which proposed to build an outfall for no more than one hundred fifty thousand dollars. The Pacific Sewerage Company offered a new plan to build filtering and drying beds at the southern terminus of the San Pedro Sewer for twenty-five thousand dollars and said it could then sell the sewage to farmers in Vernon. Although for a while it appeared that the council was seriously considering some of these plans, all were eventually rejected.

Despite continuing opposition to Eaton's plan, city officials drew up the necessary right of way for the outfall. The city negotiated with the Southern Pacific Railroad and prepared the ordinances for publication. A bond measure for $1,280,000—covering the construction of the three interceptors inside the city limits, the outfall sewer and a separate storm drain system—was placed on the ballot on August 30, 1889. The voters overwhelmingly defeated the measure. Newspaper reports noted that everyone favored the interceptors but no one wanted an ocean outfall.

In November, in an action that would be repeated several times over the next fifty years, the council appointed a special Board of Engineers to review all the options open to the city. The board consisted of Rudolf Hering, City Engineer Fred Eaton and two respected local engineers in private practice— George C. Knox and August Mayer. The board presented its report to the council the day before Christmas.

After listing several options and the cost of each, the board concluded that a Los Angeles River route should not be considered because "there existed no legal authority to pollute a water course and create and maintain a nuisance." The board approved Eaton's plan for the outfall, suggesting that the sewage could be used for irrigation during the summer months. When the board interviewed owners of property along the route to Ballona, they found that although many of the forty-five to fifty property owners were using crude sewage, these farmers were not ready to enter into contracts for receiving sewage from the city.

The engineers also recommended that the sewage should be screened before going into the ocean, with the screenings burned in a furnace. They further recommended that the outfall sewer should extend about two thousand feet from the shore into thirty-five feet of water and consist of two pipes, each three feet in diameter. As things turned out, none of the recommendations for screening, length of the outfall into the water or burning of the screened material was followed when the outfall was completed in 1894, four years after Eaton finished his second term as city engineer.

While the mayor dithered and the council hired more experts to tell it what to do, people in the city were overwhelmed by the stench of sewage gases which continued to escape from houses and sewage pipes. It became especially unbearable during times when the South Side Irrigation Company held back the floodgates to build a greater head for flooding the vegetable gardens. City health officer Granville MacGowan said that "the gases are then forced back into the city where they pour out from every manhole to the disgust and inconvenience of everyone within reach of the stench."

Sewage odors emanating from faulty house connections and from poorly ventilated and inadequately constructed older sewers continued to plague the city. Finally, in March 1890 the council placed three measures on the ballot to correct the situation. Instead of one all-inclusive measure, the council submitted three separate ones: one for $374,000 would provide for the construction of the three major interceptor sewers and additional mains within the city limits; a second measure for $527,900 would cover costs of storm drain sewers; and a third for $696,775 would fund the outfall sewer. The measure to build the interior interceptor sewers was approved, but the measure requesting money for the storm drains failed, as did that to build the outfall sewer.

By December 1890, as Eaton was finishing his second term of office, he was able to report that the Western Interceptor was half complete: one-fourth of the Southern Interceptor was finished and two sections of the Central Interceptor had been completed. That still left the problem of what to do with the sewage once it was collected. In his final annual report, Fred Eaton noted that the city still needed some kind of outfall. In desperation he suggested discharging the wastewater from the Central Interceptor to the dry bed of the Los Angeles River southeast of the city as a temporary measure, using it also as a storm-relief sewer during wet weather.

For the rest of the sewage, Eaton now proposed to carry it by way of his original route through the Centinela Ranch lands for approximately two hundred fifty thousand dollars. To achieve the savings he proposed using smaller pipes made of steel, then switching to cement at Hyde Park, continuing to Inglewood, where steel pipes could again be used for about thirty feet beyond Inglewood, where an open ditch would be cut to convey the water to the ocean. This way, he argued, the water could be used for irrigation of four thousand to six thousand acres during the dry season and in wet weather the waste could be dumped into the ocean.

Angry and anxious about the reluctance of the residents to approve an outfall, Eaton insisted that the sewage must be carried away from the city. Otherwise, the frustrated engineer said, "Los Angeles will become a breeding ground for the most dreadful diseases."

After leaving office in December 1890, Fred Eaton disconnected himself from the city's sewerage problems, pursued his other business interests and remained active in Republican politics while continuing to consult with the city on water issues. Eaton began a decade-long struggle to bring the city's water supply under the control of the municipality. He knew that the Los Angeles River could not supply enough water for the city's growth. In the summer of 1892, Eaton took the first of several trips to the Owens Valley, where he began plans for a new source of water for the expanding city.

In December 1898 Fred Eaton was elected mayor. While in office, he created a Committee of 100, which successfully promoted a bond measure to acquire the Los Angeles River water rights from the Los Angeles Water Company. A series of lawsuits followed, so it was not until 1902, two years after Eaton left office, that the city prevailed. Eaton became obsessed with the idea of finding another source of water for the city. He again traveled to the Owens Valley east of the Sierras and attempted to interest his friend William Mulholland in reaching out to the abundant waters of the Owens River for an aqueduct to Los Angeles. Mulholland initially dismissed the idea, saying that the Los Angeles River had enough water to supply future generations.

Following the extremely dry winter of 1903-1904, Mulholland agreed to visit the Owens Valley. In September he and Eaton made their now-famous trip by mule to the upper Owens Valley, where Mulholland made notes and sketches of the plan that Eaton outlined to bring water via an aqueduct to Los Angeles.

Eaton and Mulholland's friendship ended when Eaton attempted to wrest extra money for land he had acquired in the Owens Valley while acting as the city's representative. By this time, Eaton's life was in shambles. He was a heavy drinker who divorced his first wife and married a woman twenty years his junior. Fred Eaton was an argumentative visionary who had taken up his father's devotion to water, but spent his declining years in bitter recriminations against the city he had once served. Unfortunately, Eaton never recovered from his inability to profit from earlier land purchases in Inyo County and died penniless in 1935 after a series of strokes; his land was claimed by banks in lieu of payment on loans. After his death, his widow was reduced to living on charity. He had several children, none of whom were known for any civic contributions. His youngest son, depressed by his father's unsuccessful capitalistic ventures, joined the Communist Party and in 1936 was killed fighting against Franco in Spain.

Despite the breach between the two men, Mulholland always referred to Eaton as the "father of the Los Angeles Aqueduct." Nevertheless, Fred Eaton has been relegated to a footnote in history despite being credited in 1930 with having designed and laid out Pershing Square, Elysian Park, Westlake Park (before Gaylord Wilshire expanded it) and Lincoln Park. In 1868, at the early age of thirteen, he created the original design for the Plaza—today's Olvera Street.

Although the original outfall to the ocean was built by his successor, Fred Eaton should certainly be remembered for his successful efforts to municipalize the city's water supply, for his sustained interest in importing water from Northern California, for resisting efforts to send sewage to the Los Angeles River, for planning the first outfall to the ocean and for designing the first integrated sewer system for the city.

CHAPTER FOUR

Sewer to the Sea

In March 1892, two years after Fred Eaton left the city engineer's office, the voters finally approved construction of the new sewer outfall, pretty much as Eaton had originally suggested in 1887. John Henry Dockweiler, who had been elected city engineer in 1890, was serving his second full term as city engineer as he supervised the work. He was the eldest son of Henry Dockweiler, who, like Eaton's father, had come to Los Angeles after the Gold Rush of 1849. Unlike Fred, Dockweiler had a stable family environment. He grew up in his father's household, and Henry Dockweiler had put his roots firmly in Los Angeles. He was the owner of the La Fayette Hotel, a major social center in Los Angeles, and ran several other businesses. An active Republican who served on the Common Council from 1870 to 1874, Henry was a respected and prominent member of Los Angeles society.

◄ Nineteenth-century Department of Engineering staff.

Several months after taking office as city engineer, Dockweiler began planning to build the sewer to the sea. He asked the city attorney to begin preparation of condemnation suits for the right of way through Daniel Free-man's legendary Centinela Ranch, along which the sewer route would be laid. The wealthy and politically active Freeman immediately fought the proposal.

In October 1891 the council held several contentious discussions about Freeman's rights to the land and payment for the sewer right of way. In December the council approved the final route of the sewer. On the following June 9, the state court gave the city the right to his ocean frontage land where the sewer would empty into the ocean, for which the city paid $2,500. A jury hearing that month awarded him damages of ten thousand dollars for the perpetual right of way across his land through the town of Inglewood, for a distance of about twelve miles. Freeman was extremely worried about sewer gases and obnoxious odors that he was convinced would emanate from the new sewer. Dockweiler as-sured him that the sewer would be built four feet underground, would be closed and would have no manholes from which gas could escape.

On July 12, 1892 the council approved the final plans for the outfall sewer. The ordinance also established the final route of the outfall almost di-rectly southwest from the Los Angeles city boundary through the southern portion of what was then Inglewood, near Prairie Avenue. Tired of the stench, now reaching all parts of the city, residents approved a bond measure to build the outfall sewer on August 31, 1892. Because the funding was about half the amount Eaton had estimated the sewer line would cost, the new sewer would not be built of the material Eaton had specified.

Throughout his career Eaton had been a staunch opponent of the use of wood for carrying either fresh water or sewage. In a 1905 pamphlet writ-ten for an Oakland sewer-bond election, he wrote that he "had watched the construction of a wooden conduit by the Los Angeles City Water Company, which I had formerly advised against, and observed its gradual decay thereafter, rendering the conduit entirely useless in less than ten years." Dockweiler had no such qualms about using wood, and approved building almost half the new outfall with wood stave pipe.

Construction of the new outfall started in the fall of 1892. Because the bond issue provided only half the funds needed to build a sewer completely of brick, half was made of redwood stave pipe bound with iron bands, while the other half was built of brick. The work was done by five different contractors, three of whom received contracts for more than one section. When the outfall was completed on March 9, 1984, an elaborate dedication ceremony was held with upwards of three hundred people on hand. Everyone was thrilled to see the project finally concluded. As a local observer wrote in an article for the *Los Angeles Times*, "Chairman Nickell and council members Munson and Strohm have been drumming up conveyances for the past three days, and they have secured everything from a gorgeous, gilded tallyho coach to the typical decrepit one-horse shay."

A valve was turned, the sewage flowed and several more speeches were made by the mayor, council members and Dockweiler. At the dedication, Dockweiler declared, "This great and needed public improvement, the construction of the sewer has added half a million to the wealth of our city and county directly, and many millions indirectly."

When the sewer was complete, the city health officer notified all property owners to abandon and fill existing privy vaults or cesspools existing and connect with the sewer within a given number of days. From that date on, whenever homes or businesses were built, the city required owners to connect to the city sewer system whenever street sewers were available.

So, the first chapter of the city's sewer history closed with a successful completion of the Dockweiler Sewer. Within a year, however, Daniel Freeman began complaining about the state of the new sewer. In a May 1895 letter to the Mayor and council, he wrote, "Yesterday one of my mares fell through a deep hole where the sewer had caved in near Inglewood." He pointed out several cave-ins along the line of the sewer, and a few days later submitted a bill for $115.00 for the loss of his mare and for the expenses incurred in retrieving it.

In 1895 a new job was created to supervise maintenance and operation of the outfall, now called the Dockweiler Sewer. C.F. Derby was elected to the new position. Upon assuming office, he faced several challenges. Most significantly, some of the brickwork was already beginning to fall. Hydrogen sulfide gas, created by accumulated sewage backing up from the wood siphons, formed a thin layer of sulfuric acid. The acid attacked the mortar holding the bricks and eventually weakened them to the point where the loose bricks would fall from the upper portion of the sewer to the bottom. Strong sewage odors leaking from the crippled sewer line were evident along the outfall route.

To halt deterioration of the iron bands holding the wooden staves to-gether, Derby requested additional water to flush the solids through the outfall. He also wanted to replace the sealed manhole covers (which originally had been used in an effort to stem the odors) with perforated covers that would allow air into the pipes and ventilate the system. His suggestions were not accepted.

In 1897, J.H. Dockweiler asked J.B. Lippincott, a well-known local water engineer, to examine operations of the outfall. Lippincott had previously worked on several special projects for the city. In fact, Dockweiler had hired him during construction of the outfall to design a drop chamber for installation at the end of the wooden section about a mile from the beach.

When Lippincott visited the outfall that summer, he saw a huge pool of sewage eddying in the sand. In the report to the city engineer, he wrote, "No effort had been made to protect or anchor the pipe portion lying on the ocean bed. Storms have shifted the pipe about fifteen feet out of alignment, causing it to break on shore at a point about one hundred feet from the line of low tide."

This resulted in sewage rushing out of the pipe, cutting a large basin in the sand. "Section after section of the pipe would break off and drop into the basin below," Lippincott wrote. Despite efforts to correct the problem with construction of a wooden flume, the sewage whirled in an ever-growing cir-cular pool. The result was a channel of about four hundred feet from the base of the bluff to the ocean, where the sewage was deposited. So the first sewage related "brown acres" were created. In his unpublished monograph, intended for the national journal *Engineering News*, Lippincott also pointed out the poor installation of the hydrants along the outfall and remarked that "the form of the hydrants, would indeed be a pronounced success, if one wished to take a sewage bath."

In April 1898, outfall superintendent Derby again reported on the "buildup of sewer gases that were steadily destroying the mortar, brick and iron of the sewer, chambers and manholes." As he suggested several years before, he replaced the closed manhole covers with perforated ones and installed ventilat-ing stations in the outfall. In June, after the installation of ventilating stations in and around the land owned by Daniel Freeman, Freeman wrote to the council Sewer Committee in a rage:

> There has been erected on the outfall sewer east of Inglewood a structure
> for ventilating the sewer that is a nuisance and a menace to the health of

everyone in the vicinity and it must be removed. In former years the city engineer kept the sewage so diluted with excess water that no damage was done either to the health of the citizen or the sewer. Under my contract with the city, the sewer was at all times to be kept in such a state as not to allow the escape of poisonous or noxious gases.

There is no record that the council ever replied to this missive. Shortly before leaving office in December 1898, Dockweiler reported that sewage flowing into the outfall, "is, for some reason not easily explained, more concentrated than I have ever before seen emanating from a community similar to that of this city."

Frank Olmstead was elected city engineer on December 15, 1898 and served a full term of two years before retiring to private practice. Eaton (elected mayor in the same election) and members of the council became concerned about the deteriorating outfall. They again hired sanitary engineer Rudolf Hering, this time to evaluate its current conditions.

Hering was effusive, but not specific in his remedies. He suggested that the sewers be kept as clean as possible (but offered no methods to accomplish this) and to allow the greatest possible circulation of air within them. He did note that "by sealing the manholes, the foul air had to escape someplace," which it did, usually into people's homes.

Probably the most discouraging news Hering brought was that the flush tanks installed throughout the city were not operating properly. He suggested that gates be installed on them to allow the sewage to be dammed up and then released in a rush to provide the necessary velocity to clean the lines. City Engineer Olmstead included Hering's report in his annual report of 1899 but made no comment about the recommendations. Instead he warned that the council's continuing indifference to the growing problems might cause the sewer to collapse if these critical situations were not addressed. Reporting a net income of some five thousand dollars from the sale of sewage that year, Olmstead wanted the council to authorize repair and replacement of automatic flush tanks as well. If the work wasn't done, he wrote, the "the situation will result in scandal and censure."

There was no scandal or censure, but the sewers continued to disintegrate, even though automatic flush tanks were installed. To reduce the flow through the outfall, Olmstead installed a septic tank at the base of the first siphon, three and one-half miles from the western border of the city. Unfortunately, it was within a mile of the elegant ranch house of Daniel Freeman, where he complained that "the nuisance was unbearable." Even with the creation of the septic tank, the disintegration of the outfall continued.

Meanwhile, the city was enjoying the profits from its sales of sewage. Superintendent Derby was soon forced to post bond because he was handling thousands of dollars in receipts. At the close of the nineteenth century and well into the first decades of the twentieth century, sewage was commonly used as fertilizer, especially in the Southwest where water was scarce and where rain could not be counted on to irrigate crops. Sewage-laced water was applied to vegetable crops as well as to barley and wheat fields. In Los Angeles, Chinese gardeners provided most of the vegetables sold in the city from areas irrigated with sewage In 1892 Pasadena had opened a sewage farm on three hundred acres in Alhambra upon which English walnuts, pumpkins, corn, barley and wheat were planted. (The farm remained in operation for several years before the city built a treatment plant in 1914.)

In 1900, Harry F. Stafford, who had served as deputy assistant city engineer during the previous year, challenged Olmstead and won the election. Stafford made the construction of a new outfall his major goal while in office. Nuisance suits were reported with increasing regularity. When he and his staff examined the outfall, he concluded that the situation was desperately urgent.

Stafford asked for construction of a new sewer—one that would be six feet in diameter, with a continuous downgrade, and constructed entirely of masonry. Commenting on objections to the sale of sewage for irrigation from the California State Board of Health, Stafford suggested that the new sewer include a purification works. If the water were cleaner, Stafford believed, then sales of the sewage water would probably increase. Removal of some of the solids would reduce most of the objections to the noxious odors.

By 1901, the Dockweiler Sewer was emitting sewage smells along several miles of the outfall. The three-mile, inverted wood stave siphon was constantly breaking. When the sewer broke, the city released the sewage into ditches that supplied irrigation to farmers on adjacent land. Because of the offensive odors and fear that the sewage could cause widespread disease, the county health officer issued a cautionary report about the use of sewage by farmers.

In December 1901 Stafford offered a plan for a new outfall to replace the deteriorating Dockweiler Sewer. Less than a year later, a $1.5 million bond measure was approved by the voters on October 29, 1902. Some of the money was set aside for construction of storm drains and bridges, but the bulk of the bond measure, one million dollars, was earmarked for the new outfall and for sewer extensions needed throughout the rapidly expanding city.

The City Council passed an ordinance that prohibited the sale of sewage for irrigation of some fruits and vegetables, while allowing it to be used on vegetables that were cooked, "or grew in such a way that they did not come into contact with the sewage." However, the ordinance was not enforced. This was undoubtedly because the city was getting a lot of money from the sale of sewage. From 1899 through 1907, when sewage sales ended, after expenses and salaries were subtracted, profits ranged from $1,500 to $5,200 a year.

Other sewage farms throughout Southern California operated until well into the late 1940s when suburban development pushed out farmers. Many may find it appalling that much of the vegetable produce sold before World War II came from farms irrigated with raw sewage. Nevertheless, these farms did not close because of health issues. In fact, according to the authors of a major study of sanitation practices in the Los Angeles area, published in 1952 by the Haynes Foundation, "Aesthetic considerations rather than threats to the purity of the water supply or other health factors," resulted in closure of some farms and limited the opening of new ones.

It took five years to obtain voter approval for the first outfall to the ocean. Once that occurred, the outfall itself was completed in less than two years. Its replacement, the Central Outfall Sewer, was approved by voters in 1902. However, it would take five years to complete.

CHAPTER FIVE

Sewage Woes

Greed, incompetence and bad luck created a series of mishaps in the construction of the new outfall sewer. Shortly after the bond issue was approved by voters, the city found them impossible to sell. Eastern investors were not interested in buying bonds that paid three and one-half percent interest. It wasn't until December 15, 1903 that City Engineer Stafford, worried the bonds had not been sold, assured the council that once they were sold, construction would start immediately. On December 30 a consortium of ten local banks led by Farmers and Merchants National Bank bought the bonds.

While waiting for the bonds to sell, Stafford had insisted that he could not issue any request for bids to build the new sewer until he was able to hire a brick manufacturer to supply the twenty million bricks that would be needed

◁ Simons Brick Company provided several million bricks for the Central Outfall in 1905-1906.

to line the outfall. The price of bricks at the time was between six and eight dollars a thousand. To get the best possible price, Stafford insisted that the city buy the bricks, avoiding contractors as middlemen.

The brick bidders responded on May 26, 1903. Although three bidders responded, Stafford considered only two of them. The Simons Brick Company was the largest, with plants in Pasadena and in Montebello, a few miles east of downtown Los Angeles. For the past twenty years, the Simons Company, run by Reuben Simons and his three sons, Joseph, Walter and Elmer, had supplied bricks for many projects in Los Angeles and Pasadena and was on its way to becoming the leading brick company of the city.

The Simons Company bid $6.75 per thousand for bricks delivered from its new plant in Inglewood, and a penny more if the bricks came from its plant in Los Angeles. A second firm, the C. Forrester Company, also known as the Independent Brick Company, offered to provide bricks for $6.40 a thousand. Stafford decided to grant the contract to the Forrester Company.

Bricks had to be solid enough to absorb no more water than eight percent of a brick's weight or they would soften and crumble off. Walter Simons lodged a complaint protesting the choice of Forrester, which everyone knew was really a combination of four young men who had never operated or even worked for a brick plant. However, they were Republicans like Stafford, and Walter Simons was a Democrat. Although the Independent Brick Company had no bricks to submit, the owners eventually took over a brick firm near Inglewood and supplied those bricks for a test, conducted months later.

In early January 1904, Stafford received the results of the brick tests. The Independent Brick Company submitted bricks that absorbed almost ten percent water, a full two percent over the specifications. Stafford assured the council that the new firm would be able to meet the necessary specifications when it upgraded its plant in time for construction. In a clear reminder that he had always intended to give the contract to the Independent Brick Company, Stafford insisted that "the test is immaterial." Over Walter Simons' protests, Stafford awarded the brick contract on January 19, 1904 to C. Forrester for $6.40 a thousand. On February 2 Forrester put up a bond for forty thousand dollars as assurance.

Once the brick contract was signed, Stafford released his specifications for the outfall and requested bids from construction companies. The bids were opened on May 16, 1904. Local contractors bid on individual sections that would bring the total cost to $666,000, and a consortium of contractors bid $698,000 for construction of the entire outfall. Stafford wanted to keep costs down, but he was unhappy with some individual bidders and recommended rejecting bids on three of the sections. He asked the council to re-advertise. On June 7 the council rejected all bids and ordered the city clerk to re-advertise.

Meanwhile, Forrester was having financial problems. One associate was accused of fraud and quit the firm. Stafford examined the kiln at a brickworks it had recently acquired and confidently proclaimed that Forrester would be able to provide brick within the eight-percent water absorption limit. Despite the troubled financial circumstances of the company, Stafford remained committed.

On June 27, the new bids for construction of the outfall were opened and a prominent Pasadena construction company, S.J. Edwards, bid $565,000 to complete the entire outfall. As low bidder, it was awarded the contract. Within two days of its bid, citing a heavy workload in Pasadena, Edwards turned the work over to a local company, Stansbury & Powell. Although Stansbury & Powell had not submitted a bid, the company had earlier accompanied Stafford and other potential bidders on a visit to the areas where the outfall would be built. There was no objection to the turnover and Stansbury & Powell began work on July 26, 2004. Within weeks, several hundred men were at work. Stafford announced that work would be completed within eighteen months.

As work continued on the outfall, it became clear that the Independent Brick Company could not deliver proper bricks as it promised. Stafford now rejected its bricks because they remained below his specifications. Finally, in September, Stafford advertised for one million bricks from a second source, even as he insisted that eventually the Independent Brick Company would be able to produce satisfactory brick. The Los Angeles Pressed Brick Company offered to sell the city one hundred thousand bricks for ten dollars a thousand. The offer was accepted. In October Stansbury & Powell agreed to wait for more bricks. Finally, in November 1904 Stafford acknowledged the failure of the Independent Brick Company and requested bids for a new contract for more than eighteen million bricks.

Because of the delay in the delivery of bricks, the old Dockweiler Sewer continued to be a problem. Sewage was pooling along the beach because of breaks in the pipe to the ocean. While waiting for bids from the brickmakers, Stafford built a new eight-hundred-foot iron pipe and a pier to extend the outfall into the ocean.

At the same time, population of Los Angeles continued to grow. By 1904, the population was estimated to be approximately one hundred thirty thousand, creating even greater burdens on the sewer system. At the same time, more than forty thousand people were without sewers. Nevertheless, the health officer said that the newer cesspools were probably better than older sewers, so an immediate solution was not necessary. Sewage flow was running about ten million gallons per day (MGD). The decade-old, worm-eaten, soil-rotted, rust-bound outfall continued to back up during winter rains, flooding streets and cellars. The sewage-laden water would then run into storm drains that eventually emptied into Santa Monica Bay. Shortly after he completed construction of the new sewer pier, Stafford told the council that he could not provide additional sewerage for the city until a new outfall was completed.

In February 1905 the Pacific Electric Railway agreed to haul bricks from the two brick manufacturers. Los Angeles Pressed Brick would deliver nine million bricks from its yard in Santa Monica for $6.35 a thousand; Simons would deliver 8,871,700 bricks at $6.38 a thousand from its yard in Inglewood and a new yard Simons would build in Santa Monica. After the contract was signed that month, the two companies each delivered one million bricks a month.

Sometime during this period the Pacific Electric established and named the Hyperion stop—a freight stop on the beach near the current treatment plant. The land that the city had bought from Daniel Freeman had remained without a name since the Dockweiler Sewer was finished in 1894. Moses Sherman and Eli Clark, the Pacific Electric's owners, were heavy real estate investors. In the summer of 1903 Sherman and Clark purchased beachfront property between Manhattan Beach and Hermosa Beach. They called it Shakespeare Beach, hoping to attract a community of writers and literary types. The new freight stop at the beach also needed a name. Sherman, who was once a schoolteacher, may have selected the name Hyperion, Greek god and father to the sun and moon, because the pier extended into the ocean where the setting sun could be seen; or the name may have just appealed to him, appearing as it did several times in Shakespeare's plays, most notably *Hamlet*. Later, however, workers believed that Sherman and Clark were making a little dirty joke, since Hyperion was the offspring of Uranus, an unfortunate—but humorously appropriate—homonym of "urine" and "anus."

After the dry winter of 1904, winter storms in 1905 slowed work on the outfall. In March, shortly after the new brick contracts were signed, the worst storm in twenty years halted all work on the outfall sewer. Several bridges throughout the city were destroyed, as were recreational piers at Ocean Park and Playa del Rey. The San Gabriel River breached its banks and the Los Angeles River rose to its highest level in twelve years. The new sewer wharf, completed only a few months earlier, was completely washed out as well.

Engineers at that time were unaware of Santa Monica Bay's unique submarine canyons. The sea-bottom topography of the bay includes two major canyons. During winter storms, these canyons focus large waves coming in from the west, especially at the site of the new outfall, often reaching heights greater than twenty-five feet, and pier washouts were a common result.

After several portions of the new sewer in the tunnel under Inglewood were washed out, Stansbury & Powell said it would bill the city for damages. Nevertheless, by July the company had resumed work on the outfall and repaired parts of the sewer tunnel in Inglewood.

In the following months, as work continued on the new outfall, workers were beginning to organize in the city. The early decades of Los Angeles were notorious for aggressive police activities against labor organizers. *Los Angeles Times* publisher Harrison Gray Otis and his son-in-law, Harry Chandler, joined the Merchants and Manufacturers Association, a powerful anti-union group created a decade earlier, and soon became leaders in the battle against organized labor.

A struggle over work on the sewer—which employed hundreds of workers—soon ensued. Despite the suppression of the American Railway Workers Union, broken when President Grover Cleveland deployed thousands of federal troops, including a contingent that went to Los Angeles, workers in the city were actively organizing unions in the early years of the twentieth century.

In November 1905, representatives of these local unions petitioned the City Council, alleging that the contractors on the new outfall sewer were not using contract-specified materials. The work was being performed by un-skilled workers who had been actively solicited by anti-union businessmen in the area. Mayor Owen McAleer, William Mulholland (still the head of the city's Water Department), members of the City Council Sewer Committee and City Engineer Stafford visited the outfall site and declared the work adequate. Suspicions about the quality of work continued to be debated in the City Coun-cil. In December the mayor agreed to appoint an independent investigating team that eventually proclaimed the work acceptable.

Less than two months later, the only known suicide by drowning in sewage occurred in February 1906. The *Santa Monica Daily Outlook* reported on February 8 that a Miss Jennie Brown, "described as an elderly woman, was seen near Hyperion by a conductor of the Los Angeles Pacific Redondo line. Searchers found her tracks leading to the Los Angeles outfall sewer tank, but no tracks could be found running from the tank." The writer speculated that she "threw herself into the tank and was carried through the outfall pipe into the sea." This was a local story not carried in the metropolitan newspapers. Hers was the first and least-publicized suicide story attached to the outfall.

On the political front, a coalition of "good government" groups led by Dr. John Randolph Haynes led a reform movement in the city in 1902. These men, who considered themselves Progressives, created the nation's first legisla-tion establishing the initiative, recall and referendum process that allowed citi-zens to participate directly in writing legislation and to recall political figures. Progressivism changed many of the city's corrupt political practices and created a Board of Civil Service Commissioners that established civil service protection for city employees. In 1905, the Progressives succeeded in enacting a charter amendment creating a Board of Public Works to replace the City Council's public works commission, which had little public accountability.

The new board consisted of three members, each of whom would be paid an annual salary of three hundred dollars. The mayor appointed the three members but the new appointees needed approval of the City Council. Board members would be the final arbiters in awarding contracts for all city work. Council approval would not be required. The board, in turn, would now ap-point the city engineer, who had been elected since the 1889 charter change. The new Board of Public Works, seated in March 1906, was suspicious of the contractors who had been hired by the old City Council. In an effort to be fiscally responsible, these new board members were not inclined to approve additional funds requested from Stansbury & Powell.

In April 1906 Stansbury & Powell again encountered water spilling into the tunnel beneath Inglewood. The contractor temporarily halted work on June 17. To avoid liability, the firm complained that the city engineer had laid out the tunnel so that it would go seventy-five feet under the Inglewood hills through the water table, which caused water to constantly flow into the tunnel. What's more, it claimed that Stafford was accepting brick that did not meet specifications.

According to most reports, the new Board of Public Works that had been seated only a month earlier sat idly by, so Stansbury & Powell complained to the City Council. Charging that regular pumps would not be enough to remove the water, the contractor claimed that shields would have to be built and pushed into the quicksand to hold it back before they could continue work. The firm also told Stafford directly that it could no longer work according to his specifications.

The water department's William Mulholland defended Stafford. He insisted that the work could be done with pumps that would create cones of depression along the line while the masonry was constructed. City Engineer Stafford insisted that the contractor knew the water was there when it made its bid. The contractor wanted a change in the specifications to allow the sewer line to be laid above the water level. This would mean installing pumping plants to raise the sewage; the sewer would not be on a gravity flow. The city officials refused to abandon the gravity-flow plan of the outfall. With the backing of Mulholland, the Board of Public Works insisted that the work could be done without the change.

On July 15, observers could see more than one hundred and fifty men and fifty mule teams working on two sections of the outfall. One day later, teamsters were taking the stock away to city corrals. On July 17, the mayor and members of the Board of Public Works visited the tunnel and found no men at work. They soon learned that workers had been paid off and dismissed.

The city attorney announced that the city would hold the contractor to its promise and that it would forfeit bond. The contractor formally served notice on the city that it could not complete the work. On August 2 Stansbury

& Powell wrote the city and again claimed that the work stoppage was not its fault. The city attorney gave the firm five days to resume work.

That same day, Harry Stafford was found unconscious in the locked bedroom of his gas-filled house. The windows had been shut. He died the next day. All the news stories that reported his unconscious condition on the day of the incident remarked on his unhappiness about the delays in completion of the outfall and the walkout by the contractor. Although two physicians agreed that it was an accidental death, rumors of suicide circulated widely. Three local newspapers, The *Express*, The *Los Angeles Times* and The *Examiner*, carried stories about Stafford on their front pages for two days. Each printed the speculation that Stafford had killed himself in despair over the outfall debacle. No coroner's jury was called nor was an inquest held. The public and city officials accepted the verdict of accidental death. The gas that killed Stafford came from fumes emitted from a gas jet of what was described as an instantaneous heater. The pilot light had not been lit. The vent pipe was found to be broken, and later it was speculated that an explosion had perhaps caused the pilot light to be extinguished. Yet the heater itself remained intact. Flags were flown at half-mast on public buildings "over the death of this pioneer of Southern California." No flags had flown for poor Jennie Brown, who died in near obscurity. Stafford's wife, who had been in Catalina with their two children, insisted that his death was an accident. She claimed that the same heater had malfunctioned two weeks earlier but had not been repaired. However, it is likely that Stafford, an experienced engineer, was aware of the faulty gas heater and on that hot August day he ended his own life.

In the weeks and months before Stafford's death, the daily newspapers were filled with stories raising questions about work being done and money spent on the new outfall sewer. Headlines such as "Los Angeles' Money-Eating Sewers," "Outfall Bubble Bursts," and "All Seek to Avoid the Responsibility," led pages-long articles on blunders and cost overruns of the outfall construction. One week after Stafford's death, an inquiry was opened into his approval of bills submitted by Stansbury & Powell. The investigation continued for a few weeks, but was abandoned without explanation.

Homer Hamlin, who had long worked in the city engineer's department as an assistant deputy city engineer, was appointed the new city engineer. He quickly assumed the task of completing the unfinished work on the outfall—25,827 feet of the total length of approximately 64,378 feet—mostly in a tunnel near Inglewood. It would be December before that work could begin.

Meanwhile, Hamlin reported that the old outfall had reached its capacity and in November the city engineer ordered that sewage run over the 406 acres to temporary ditches. Bids for the new outfall were advertised and received, but Hamlin argued that the city could construct the remainder of the outfall with city forces instead, for less than the lowest bid received. The board agreed and instructed him to proceed. Wells were dug, lumber was ordered and the payroll increased several-fold as the city established its own boarding camps for the new temporary employees.

To prevent settling and development of cracks, and to permit waterproofing, Hamlin substituted the inverted brick originally called for in the tunnel with a wide concrete base lined with a ring of bricks. Before work began, part of the long tunnel near Inglewood collapsed, killing four mules that were being used in a farm field.

In the years it took to complete the new outfall, the old Dockweiler outfall continued to plague the city. Complaints about the sale of sewage, the growing threat of lawsuits from the odors, and California State Board of Health concerns initially caused the Board of Public Works to halt the sale of the waste material in August 1906. The board was forced to rescind this order within weeks when the city engineer informed the board that the excess sewage flow would inundate city streets if it was not released onto land surrounding the outfall. In October, Hamlin constructed two interceptor sewers connecting the old sewer to recently completed sections of the new Central Outfall to provide some relief to the old outfall.

During that winter's storms, sewage was allowed to escape over private property to alleviate pressure in the siphons. The original contract with South Side Irrigation had expired, so the City Council signed a contract with a new firm that owned land east and south of the Redondo Railway on the La Cienega Rancho. The new contractor would take all overflow sewage and distribute it over its lands free of charge, for at least two irrigating seasons, unless the new outfall was to be completed before then. Many more claims of damages were filed during these months before completion of the new sewer.

Day workers began construction of the tunnel on February 18, 1907. On September 29, 1907, the new outfall, an inverted horseshoe made of concrete and lined with brick and mortar on the upper invert, was completed. The new sewer was intended to run half-full with a capacity of more than thirty-two million gallons per day. As long as the flow remained at that level, there would be sufficient air space above the flowing sewage to which would allow rapid oxidation. Manholes were placed at intervals of every five thousand feet, except in the tunnel portion. When the sewer was finally opened for use, the old Dockweiler outfall was abandoned and sewage sales ceased. A deficit of $321,500 was absorbed by the city without recourse to a new bond measure. The new outfall sewer, which would be known as the Central Outfall, was put into operation on November 1, 1907.

CHAPTER SIX

More Sewage Misery

In March 1908, following completion of the new Central Outfall, City Engineer Hamlin replaced the old cast iron pipe connecting the outfall to the ocean with a forty-two-inch concrete pipe lined with brick. By June 1909, Hamlin reported that the sewer was running at eighty percent capacity, considerably higher than planned.

By 1910, the city had annexed more than seventy additional square miles, including the 1906 annexation of the "shoestring district," a long narrow corridor that connected Wilmington and San Pedro to the city, which allowed Los Angeles to develop the harbor in San Pedro. City sewers extended for a total of 529 miles. From 1900 to 1910, the population had tripled from 102,000 to 320,000, and many homes were without sewer connections. Sewage from these areas was still discharged into open ditches and storm drains.

In January 1913, members of the California State Board of Health, then responsible for regulating sewage facilities statewide, visited Hyperion and notified the City Council that the conditions there needed to be abated. Shortly after the California State Board of Health members' visit, the rain came down on the plains of Southern California. From January through February, 23.85 inches of rain fell, more rain than the city had experienced since 1892. Sewers overflowed throughout the central and southwestern parts of the city.

In February the Board of Public Works instructed the city engineer to prepare plans for a new outfall. In December, Hamlin presented a map outlining a new outfall, the North Outfall Sewer, that would also provide sewage service to additional annexed territory. He estimated that the outfall would cost a million dollars. An additional two hundred thousand dollars would be needed for a purification plant. This was the first time city engineers began planning to treat the sewage. During this period, Hamlin relied heavily on his sanitary engineer Willis T. Knowlton, a dedicated civil-service employee who was responsible for supervising work on all the city sewers.

Knowlton suggested that the city build a test plant before going to full-scale treatment. He wanted to install Imhoff tanks (a method that allowed solids to settle, after which they would be removed for burial or other disposal). The clearer effluent would then be discharged through the outfall at Hyperion. The City Council took no action on his report, presented under Hamlin's auspices.

In 1914, Knowlton attempted to get money from the City Council to install the Imhoff system. However, despite the relatively low anticipated expense of forty thousand dollars, the council refused to approve his plans and turned down his request for a bond measure. In January 1916, due to an overwhelming rush of water from yet another big storm, a seventy-five-foot-long break opened up near Inglewood and sewage spilled out onto the adjacent land. The break was temporarily patched with hundreds of bags of sand.

One month later Hamlin reported additional damage to the line and spent fifty thousand dollars for permanent repairs. Following the completion of the Los Angeles Aqueduct in 1913, the city annexed huge tracts of land in the San Fernando Valley in 1915. By the summer of 1916, annexations of Palms and Westgate, an area south of the San Fernando Valley that extended south to the city of Santa Monica, brought the average daily sewage flow to 42 MGD, well above the capacity of the Central Outfall. In June the council finally agreed to a sewage bond measure for $1,750,000.

Despite the approval of all the local newspapers, the measure was defeated. Smug self-interest and lack of public spirit dominated the city, as voters rejected three other bond measures calling for construction of additional street tunnels and fire stations, as well as an ordinance that would have allowed the city to sell Owens Valley Aqueduct water to other cities and territories. Although the city, at approximately 342 square miles, was now the largest municipality in the United States, the far-flung nature of its suburbs worked against any civic consensus on sewer construction, especially in a time when the majority of voters were registered Republicans and known for their fiscal conservatism.

For the most part, the sewage problem was confined to the southern parts of the city. Although winter storms would unleash sewage spills into city streets, voters were uninterested in a tax increase to solve the problem.

The Central Outfall sewer had long since reached capacity, and the ocean discharge pipe had sprung a few leaks as well—discharging even more sewage onto the beach and into the surf zone. During the summer of 1916 the California State Board of Health repeated demands that the city correct the problems at Hyperion. The city was unable to comply because the demands were moot without a bond issue to fund the construction of a new outfall and a treatment plant. In August, at the behest of the California State Board of Health, the county grand jury began hearings. "An unsanitary and unsafe condition," Knowlton acknowledged in his annual report, "existed between Belleview Avenue [currently Imperial Highway] and the shore end of the sewer pier." The California State Board of Health, in vain, demanded that the city correct the situation.

The situation had become so critical that Knowlton feared that the Central Outfall would collapse at the ocean end. To save the outfall Knowlton built a new pipe that intercepted the sewage at Belleview Avenue (which is now Imperial Highway). Since no bond funds were required for this project, work began in September 1916. More than 3,700 feet of tunneling was required for this new interceptor which, when finished in March 1917, relieved the overflowing Central Outfall, which also remained in operation. The new Belleview Outfall was connected to a new pier that extended three hundred and fifty feet beyond the high-tide line and was north of the original outfall pier. The extremely offensive odors at the beachfront near the old pier were partially eliminated. Despite the extensive raw sewage surrounding the Belleview Outfall, beachgoers continued to sunbathe and swim in the polluted waters at Venice and Manhattan Beach.

In May 1917 Knowlton revised his bond measure sewer plan of 1916 and submitted it to the City Council. He planned to construct only 12.6 miles of new sewer to connect with the Central Outfall at Exposition Boulevard and Arlington Avenue. The line would go west for about four and one-half miles to the foot of the hills east of Culver City; then south about three and one-half miles along the base of the hills to a point about one-half mile north of Inglewood. The sewer would then go in a direct line in a tunnel to Hyperion. The bond measure was for a total of $1,750,000.

The timing could not have been worse. For months the newspapers had been filled with World War I horror stories. In April, the United States declared war on Germany. As voters lined up to vote in the June election, polling places were also used to register men signing up to fight the German menace.

The *Times* thought it inappropriate for the city to be requesting these funds "because of the uncertainty of financial affairs." The newspaper's stewardship was by this time firmly in the hands of Otis's son-in-law, Harry Chandler, and it was as powerful in selecting political candidates as it had been in Otis' time. Mayoral candidates and council candidates needed the backing of the newspaper to win office. Support from the *Times* could not guarantee victory, but lack of support would often assure defeat. When the newspaper editorialized, in the days before the June 5 election, "The city is getting along nicely under present conditions," voters followed its advice and overwhelmingly defeated the sewer-bond measure.

On the same 1917 ballot, two additional territories, Owensmouth (in the west San Fernando Valley) and a coastal area (Westchester and Playa del Rey) were annexed by an overwhelming vote. This brought the total city land area to slightly more than 351 square miles. Meanwhile, Standard Oil incorporated its property as the City of El Segundo several months before the June 5 election, creating both a permanent barrier to Los Angeles expansion plans and an enmity toward Hyperion among its residents that did not abate until well into the twenty-first century.

Meanwhile, City Engineer Hamlin was not getting along with the Board of Public Works, which was unhappy with his refusal to support the work of contractors he believed to be unsatisfactory. In a report to the Board of Public Works in January, a City Council-appointed Bureau of Efficiency sharply criticized his management of his office. The efficiency board alleged that the city engineer's office was the largest department in the city organization, with about 1500 employees, but Hamlin's "plan of organization is seriously defective, with sixty different chiefs of divisions and foremen reporting through the chief deputy city engineer." They argued that it was "impossible for the latter to know accurately the work being done," and claimed that no accurate records were kept of the activities of the individual employees.

Hamlin, who had been working for the city for nineteen years and was lauded for his work when he had successfully completed the Central Outfall after contractors walked off the job, felt he was being unfairly attacked. He refused to respond to the report because, he said, "it would be inappropriate." Shortly before the June election, mindful of the increasing pressure against him, Hamlin sharply criticized Knowlton, who had appeared before a public meeting convened by the Municipal League. The league had been formed in 1901 during the ascendancy of the Progressives and Good Government forces. Knowlton's defense of the improvements planned for Hyperion were reprinted in the League's monthly publication, the *Bulletin*. By this time, Knowlton was a well-known and highly regarded sanitary engineer.

In July 1917, the Board fired City Engineer Homer Hamlin. After he was fired, several local newspapers reported widespread corruption, kickbacks and inferior work by the so-called "paving trust" in the Department of Engineering. Assistant City Engineer Andrew Hansen was appointed city engineer following Hamlin's ouster.

During the years of World War I, the city did not attempt to pass another sewer bond. Unfortunately, the old Dockweiler Sewer was in bad shape. The sewer, Knowlton noted in June,

> . . . shows that an extensive deterioration of the masonry lining had occurred, the bricks and mortar being badly corroded, and the strength of the masonry work being impaired . . . On account, however, of the solid matter still being deposited on the beach, the necessity for the construction of the proposed treatment plant still remains, the estimated cost of which is approximately $1,200,000 at the present time.

After the war, in 1919, voters did approve spending $135,000, by a four-to-one vote, to construct all outfall sewers, main sewers, pumping plants and treatment plants for the collection, disposal and purification of sewage. City Engineer Hansen, however, did not request another ballot measure to

build the North Outfall and a treatment plant at Hyperion. Knowlton continued to work on plans for sewage treatment facilities and to provide basic sewer
services for the city.

In 1920, Knowlton decided that since the bond measures of 1916 and
1917 failed, the city should install some type of treatment at Hyperion that
would remove all floating solids for less money. Meanwhile, El Segundo attempted to stop Los Angeles from discharging sewage at Hyperion. Trying to
stop the flow of sewage was an effort to match the labors of Sisyphus.

The city's engineers were convinced of the need for a second additional major outfall. The public, however, was unconvinced. Those interested
in the subject at all recalled the earlier profitable sales of sewage and believed
that there must be some way to treat the sewage and make it useful for irrigation and/or fertilization. They could not understand the need for an additional
capacity line, especially when the sewers were out of sight and not backing up
into people's yards or homes.

Writing in the February 1921 issue of *Engineering News-Record*,
Knowlton laid out the city's new sewerage plans. The North Outfall Sewer,
he wrote, was designed to serve a future population of two million, a figure
officials believed would be reached thirty years later, when additional sewer
lines could be added. A branch sewer, tributary to the North Outfall, would
intercept all sewage coming from between Jefferson and Manchester, west of
the Los Angeles River, while the areas of Palms, Sawtelle, Hollywood, Beverly
Hills and South Hollywood would be connected to the new outfall. An East
Los Angeles Interceptor was also planned and could be extended to provide an
outlet for sewage coming from the San Fernando Valley. Knowlton said that an
additional one million people could be served by the new outfall.

John Allen Griffin presented extensive reports to the Board of Public
Works on plans for the North Outfall Sewer in late September 1920. The
proposed cost for the outfall and screening plant was a hefty $12,850,000, five
times the cost of proposals in 1916 and 1917, but this new plan was designed
to provide ample capacity for an expected 1950 population of three million.

Despite intense effort by city officials and a widespread publicity campaign to gain approval for the new sewerage facilities, voters remained recalcitrant. Shortly after an advisory committee of local engineers reviewed the
plans of the city engineer, opposition to the proposed outfall began to build.

Much of the opposition was based on the idea that it would be better to reclaim the water and use it for irrigation, while simultaneously recovering the sludge, drying it and using it for fertilizer. Since sewage farms were still operating at a profit during this period, some citizens wanted to utilize the sewage rather than discharge it into the ocean. Several sewage farms—Pasadena. Covina, Pomona and San Bernardino, for example—were operating at the time and were either providing sewage water for irrigation or were directly applying the sewage to a variety of food crops. In April 1921 Griffin drew up a map of areas that could be used for sewage irrigation—all outside city limits, but within the county.

Reuben Borough, a reform journalist and active political liberal, was among the leaders of an anti-North Outfall movement. Borough wrote in an article in the *Municipal League Bulletin*, "the whole scheme of outfall sewage disposal is preposterous, most archaic and unsanitary, as well as economically wasteful." He wanted the city to install treatment facilities—either activated waste, or Imhoff tanks, as Knowlton had earlier suggested.

An activated waste plant removes solids from the wastewater, then re-introduces some of the solids, now called sludge, to the cleaned water, which is then aerated in large holding tanks. Air is pumped into the tanks so that bacteria in the water can feed on the sludge. The solids grown in this second stage are called activated waste and are removed in a final settling process, and then pumped to closed tanks where the sludge is "digested," creating gas that can be recovered to use for power. Borough argued that the city could recover the sludge for sale as fertilizer, while discharging a more purified effluent to the ocean.

Responding to the criticism, Knowlton and Griffin claimed that if the city adopted the activated sludge process, it would cost $1,490,000 annually. Knowlton was convinced that properly installed screens would keep all float-ing material from going into the ocean. The plan was to discharge the effluent through a concrete pipe twenty feet below mean high tide, two thousand feet from the shore. The city appointed a three-member commission—William Mulholland of the Bureau of Water Works and Supply, Harvard Engineering Professor George C. Whipple and George W. Fuller from New York City—to study the plan. They concurred with Knowlton's conclusion that an activated waste sewage treatment plant would be too costly. Moreover, Knowlton noted, the California State Board of Health favored the screens over settling tanks, for being less difficult to maintain than an activated waste plant.

On May 18, 1921, less than a month before the election, on its front page the *Los Angeles Times* promoted a sewage treatment proposal unlike any offered previously. A Mr. C.D. Crouch outlined a plan to build a sewage plant that he claimed would screen out the solids and discharge the clarified water into the Los Angeles River or into inexpensive open ditches. After a brief discussion, the council tabled the matter. There was no political support to send sewage, no matter how clarified, into ditches or the Los Angeles River.

The next day the newspaper editorialized that there was something shifty about the behavior of Griffin in hiding his outfall plan for so long. Then on June 4, one day before the general election, citing widespread general opposition to the sewer measure, the *Times* editorialized against the measure. The sewer-bond measure was defeated by an almost two-to-one margin. Other municipal improvement measures, including waterworks facilities, fire and police department buildings, fire hydrants, construction of a central library and several branches, as well as extensive harbor improvements, were overwhelmingly approved.

Knowlton predicted that if no new outfall was built, there would be sewage in the streets by 1924. He was off by three years. There was sewage in the streets by the winter of 1921, six months after the bond measure failed.

Immediately after the devastating defeat of the sewer-bond measure in June, City Engineer Griffin and sewage engineer Knowlton toured the East Coast, visiting sewage facilities in several cities. On June 23, Griffin proposed an alternative to the North Outfall: build an activated sludge treatment plant in the southeast part of the city and another plant in the southwest part of the city, diverting approximately 31 MGD, which would provide relief for about three years. The solids could be sold or buried. The clarified liquid could either be reused for irrigating farms or discharged into the Los Angeles River. Griffin estimated the total cost for construction, operation and maintenance of this system, without the new North Outfall, at $4,150,000, eight million dollars less than the failed 1921 bond measure.

Two months later Griffin changed his mind once again. He now suggested building an interceptor sewer to serve the southwest part of the city, installing two activated sludge plants recommended earlier, constructing an eastside intercepting sewer and a screening plant at Hyperion with a mile-long submarine outfall. All of which, Griffin declared, would cost $3,450,000, seven hundred thousand dollars less than his earlier estimate.

In December 1921 the Los Angeles Chamber of Commerce supported the new bond measure as outlined by Griffin. One lone voice, that of the crusading journalist Reuben Borough, opposed the North Outfall plan in the pages of the the *Los Angeles Record* because he wanted the city to install a modern activated waste plant similar to one that had recently been installed in Pasadena. Again, calling the whole scheme preposterous, he attempted to create interest in building a plant that could use the sludge as fertilizer, similar to what Milwaukee had done. That city was then producing one hundred tons a day of Milorganite, a product still in demand in the twenty-first century. Circulation for *The Record* was nowhere near that of the other major dailies, and Borough's demands for a modern plant were ignored by Knowlton and Griffin.

A few weeks after the chamber supported the bond measure, the skies opened up and torrential rains once again deluged the city. The arid desert plain experienced the heaviest downpour in twenty years, six inches more than the previous year. The winter storms of 1921-22 devastated the city's sewer system. The storms caused overflows throughout the city, flooded thoroughfares, popped manholes and produced local geysers that erupted on a regular basis, causing locals to dub them *Old Faithful*. Streets throughout the city were flooded with storm water and sewage, as two-hundred-pound manhole covers were lifted and geysers of sewage spewed into the air. The sewage-laden water rose to fill basements of many homes, especially in the southwest part of the city. Once the rain stopped, the sewage water flowed into the storm drains where it was carried out to the ocean.

Meanwhile, film production significantly increased the flow to city sewers and open storm drains. The motion picture industry, was by this time one of the largest industries in the United States. With an annual payroll of twenty million dollars in 1915 and thousands of employees, the industry and its extensive use of water in processing film contributed more than a little to overflowing sewers. In 1922, Knowlton reported, "the moving picture industry has created a condition without precedent in this or any other city in the United States." The industry discharged, at times, several cubic feet of water per second, causing sewage to back up and flow through the streets near the intersection of Millrace and Wilcox Avenues in Hollywood. A bypass sewer was constructed, and then two additional sewers were built bringing the sewage into another part of the system.

By February of 1922 sewage was backing up all over the city. In an attempt to protect the Central Outfall from breaking, the city engineer diverted sewage from one of the mains into an open ditch alongside the Southern Pacific/Pacific Electric Inglewood Line. The sewage then flowed into Ballona Creek. The city asked for a permit from the California State Department of Health to screen and chlorinate the flow. The permit was denied. Culver City, Venice and Playa del Rey sued and the city was given until October 1922 to correct the problem. It would be another year before the problem was solved with construction of a temporary outfall paralleling Ballona Creek to the shoreline and then south to the Belleview Outfall built by Knowlton in 1917.

In late spring, a new North Outfall measure was proposed. Despite continued opposition by the *Los Angeles Record* deploring the "lack of ability, foresight and comprehension of our city government and the incompetence of our sanitary engineering department," a much larger effort by business interests caught the attention of a burgeoning, prosperous community, which was now willing to spend some money to halt the sewage woes, even if the solution was not perfect.

Angelenos had not forgotten the winter downpours and sewage outfall overflows. Although the Los Angeles County Farm Bureau, Reuben Borough and the Municipal League continued their opposition to the North Outfall and screening plant proposal, the *Times* made a 180-degree turn and now endorsed a twelve-million-dollar bond measure.

On August 20, several days before a special election for the sewer bond was held, the newspaper berated citizens for their lack of serious thought about the city's "primitive system." The *Times* produced several columns of support for a plan which was little different from the plan floated the previous year. Showing photos of "Old Faithful" geysers and children playing in the sewage-laden streets, the newspaper endorsed Griffin's plans. A speakers' bureau was formed, flyers were sent home with schoolchildren, campaign stickers were inserted in utility bills and fact sheets were inserted into circulating library books.

Appropriately, the motion picture industry produced "The Film With an Odor," that described the plans for a new outfall and a treatment plant, as well as providing graphic examples of sewage in the streets. The movie was distributed to more than thirty-five theaters and was shown at a mass meeting at the Hollywood Bowl the Friday before the election.

After the barrage of publicity and the overwhelming civic leadership support for the bond measure (not to mention the sewage overflows of the previous winter) on August 29, 1922 the bond measure passed—67,959 to 14,671. Work on the new outfall began the following year. Engineer Griffin promised, "no solid matter larger than a head of a pin will find its way into the ocean." Little did he know. Sewage problems would continue to plague the Southland for decades to come, and millions of gallons of sewage would ensure an ocean of brown acres that would provide nutrients for abundant marine life well into the 1950s.

CHAPTER SEVEN

Dilution Is the Solution to Pollution

L os Angeles was the quintessential city of the Roaring Twenties. Embod-
ied with a spirit and exuberance previously unseen in America, it was
a city of unbridled growth, greed and corruption, with an exaggerated
belief in its destiny. A city of strangers, Los Angeles literally exploded with
newcomers in the first years of the decade. By 1925 the population swelled
to more than one and a quarter million. As New York was the melting pot for
Europeans, Los Angeles was the melting pot for Americans.

"Once your middlewestern banker or farmer has made his pile,"
wrote Carey McWilliams in 1927, "he invariably longs for distant social fields
to conquer and moved by the urge of gigantic inferiority complex he migrates
to Los Angeles." The city quickly absorbed a million and one-half visitors a
year. In 1920, the city had more automobiles than any other city in the nation

◁ **Section of the North Outfall Sewer. 1924**

and by 1923 the Port of Los Angeles became the second-busiest port in the United States.

In this prosperous city at the edge of history, City Engineer John Alden Griffin appeared to fit right in. Born in Topeka, Kansas, he was an eighth-generation direct descendant of John Alden of the original Plymouth Colony. Educated as a structural engineer in Chicago, Griffin joined the engineering department in 1905 when he was twenty-five and was appointed assistant city engineer three years later. When City Engineer Hansen died in 1920 after a brief illness, Griffin was appointed city engineer.

Although all engineers see themselves as "problem solvers," some are visionary. Fred Eaton was a visionary who designed the first sewer system for the city. He fought to discharge the sewage twelve miles from the city limits instead of into the Los Angeles River as many had wanted. He was also the initiating architect for the construction of the Los Angeles Aqueduct, which brought water from the Owens Valley to Los Angeles and ensured the future growth of the city. Others, like William Mulholland, are charismatic natural leaders who leave legacies that enshrine their names in historical memory forever. Most civil engineers, however, are trained technicians who design and build our roads, bridges, highways and sewers—creating civic achievements that often outlive the memory of their names and their lives. Griffin was one of these men. His lack of charisma and innate obstinacy eventually led to his downfall.

Shortly after the bond measure was approved in the summer of 1922, Griffin assured the Board of Public Works and the City Council that he was hard at work developing specifications for the new outfall and soliciting and evaluating bids for its construction. Meanwhile, the rapidly growing city needed immediate relief sewers and other infrastructure improvements.

Sanitary engineer Willis T. Knowlton could barely keep up with the increasing demands on his staff. In a desperate effort to avert repetition of the disastrous sewage overflows caused by the winter floods of 1921, he completed two new sections of the temporary Ballona Creek outfall, which he hoped would alleviate the situation until the North Outfall could be completed. The added sewage only resulted in more complaints from beach communities north and south of the outfall.

Annexation, the backbone and creator of the sprawling city, brought additional sewage woes to the residents of Los Angeles during the 1920s. Los Angeles became a city of suburbs and urban sprawl. The city expanded to 441 square miles. Cesspools in recently annexed areas of the city were contaminating adjacent water tables. An area east of the current community of Windsor Hills, near Vernon and Fifth Avenues, for example, used a private sewer system having an outlet in the Angeles Mesa storm ditch. When the city annexed the land, little more than a square mile, it inherited a health hazard that created a new problem of stench and sewage in the small community. In the southwest sections of the city, many people built their own sewers that ended at vacant lots where sewage floated in pools. To offset the odors and unsightly debris, the city purchased additional property so engineers could build a pumping plant, which then sent the sewage from the private system into the Central Outfall sewer.

Several miles south of the city, fish canneries, springing up in the wake of the booming harbor at San Pedro, were polluting the harbor at an alarming rate. The city built a screening plant in Wilmington in 1917, but was unable to put in a pumping plant until 1923. Griffin was overwhelmed by requests for street improvements in addition to the need to develop plans for the North Outfall. He appeared regularly before the Board of Public Works, requesting additional staff and money to meet the increasing demands on the Department of Engineering, but was unsuccessful.

Griffin ordered a capacity survey of existing sewers. Completed within weeks, the survey concluded that many sewers were so badly overtaxed during the hours of maximum flow that they caused heavy deposits of solids, clogging sewer lines. Escaping sewer gas created foul odors throughout many communities. As a result, the engineers immediately developed plans for relief sewers in the most overburdened areas. Despite these efforts, residents continued to complain and city health commissioners reported on numerous occasions about the sewage stench occurring throughout the city.

In March 1923 the city insisted that the motion picture industry pay ten thousand dollars for "sanitary commercial waste and sewer relief in Hollywood." The studios complied, but insisted on constructing the new sewers themselves. Meanwhile, frustrated residents of Venice sued the city because sewage from Ballona Creek was flooding the Venice canals. On June 22, Judge J. Perry Wood issued an injunction against the city and set a deadline of September for the city to cease discharging raw sewage to Ballona Creek. Several months later, on December 3, 1923, as sewage continued to flow into Ballona, Judge Wood found the city engineer and three members of the Board of Public Works in contempt of court for failing to meet the September deadline. They were each fined two hundred and fifty dollars.

Despite repeated announcements to the Board of Public Works and notices to the public that progress was being made in plans for the new outfall, Griffin was unable to work on these plans. He appeared repeatedly before the Board of Public Works and requested additional staff and funds. Although the board authorized new positions, the budget was not increased and Griffin was unable to hire additional engineers.

In an unusual move for a politically appointed official, Griffin took his complaints to the public. In August 1922 he addressed an automobile association and spoke bitterly of his inability to hire qualified men to work on city projects. His remarks were printed in full in the *Los Angeles Times* on September 22. In the lengthy article, he alleged that he was working with the same workforce that existed in 1913. He complained about the low salary he was forced to offer prospective engineering employees and asked that a loan be given to the city to hire additional engineers. The Board of Public Works was not pleased.

Within a few days, the board responded with an ultimatum to Griffin. He was "to give no public interviews or statements upon any policies of the board or with respect to plans and specifications for the construction of municipal work." In a letter to the board, Griffin promised to "conform to the letter of the said resolution," and promised to "direct those under me to refrain from giving any information on any project which has not first been adopted by your honorable board." His letter did not completely mollify the board. Within a few days, the board insisted that he give it detailed plans and specifications for the new outfall. Griffin failed to deliver what the board wanted.

On April 16, 1923, worried that neither design nor specifications for the new outfall had been provided, the Board of Public Works hired Harvey Van Norman to advise it on the construction of the new outfall. Van Norman's reputation preceded him. As chief assistant to William Mulholland at the Bureau of Water Works and Supply (predecessor to the current Department of Water and Power) he had supervised construction of several sections of the Los Angeles Aqueduct and had been acclaimed for his work on that project.

Van Norman was hired to report directly to the Board of Public Works and was given full responsibility for building the outfall and screening plant at Hyperion. Two days later, his temporary position was made exempt from civil service. In addition to his ten-thousand-dollar salary, Van Norman requested and received approval for the purchase of several cars for his and his staff's use. He was able to hire additional qualified engineers because the board now approved salary increases for those engineers hired to work on the outfall. The board also agreed to allow Van Norman to set up his office on Broadway, several blocks away from the city engineer's office in City Hall. Knowlton was assigned to work for him and Van Norman moved quickly to begin work on the outfall.

He immediately decided to abandon the costly and time-consuming tunnel planned through the hills east of Playa del Rey. Instead, he laid the pipe in open trenches along the brow of the hills before tunneling south to Hyperion. Within eight months of beginning work on the outfall, Van Norman reported to the Board that the cost of the outfall would be less than originally anticipated, saving the city five million dollars.

While Van Norman was working on the North Outfall, Willis Knowlton was in charge of building two screening plants at Hyperion. Large screens would remove the solids, which would then be buried in the sand dunes surrounding the two plants. The south screening plant would take sewage from the Central Outfall, while the north screening plant would connect to the new North Outfall Sewer.

The south screening plant was completed in June 1924. It was then connected to the Belleview Outfall, which had earlier been constructed to provide relief to the Central Outfall. Eight eight-foot-square drum screens were installed inside the building with a capacity of approximately eighty million gallons per day.

The north screening plant was built facing the ocean, with five fourteen-foot-by-twelve-foot drums and space to add five more drum screens that would allow the plant to have an eventual capacity of 230 MGD. Screen slots were spaced one-sixteenth of an inch apart. This plant was placed in operation on April 13, 1925.

Both screening plants had a somewhat curious design. With open arches instead of windows, arched doorways, stucco walls over a frame of reinforced concrete and brick, and pitched roofs constructed of steel trusses, the building resembled the Spanish missions of the Franciscan colonial period of the eighteenth century. Even the channels carrying incoming sewage led to elaborately decorated tile facings.

In the architectural history book *Los Angeles: The End of the Rainbow*, Merry Ovnick described the cultural and economic history of the city. About the 1920s, she wrote that "Los Angeles was transformed into a wonderland of French Normandy castles, Middle Eastern harems, Egyptian temples, Tudor manors and even Hopi pueblos. Most of all there were the Spanish haciendas, stage sets for dream lives set in distant times and places that seemed more imaginative, adventurous, romantic or exotic than the present reality . . . There existed an ever-present incongruity, an impression of unreality." The sewage-screening plants were just another stage set to romanticize the treatment and disposal of millions of gallons of raw sewage. The Mission Style plants were torn down in the late 1940s to make way for a new modern treatment plant. The only evidence of their existence is in the few photographs taken by city engineers and preserved at the city archives.

The two screening plants were connected to the new submarine effluent outfall, which was completed in November of 1924. The submarine outfall was constructed of precast reinforced concrete pipe with an internal diameter of eighty-four inches for the first 5,145 feet from shore at a depth of about fifty feet. Because of the sandy seabed, the engineers placed the pipe in a trench dug below the ocean bottom. At that point the submarine outfall divided into two sixty-inch diameter branches at right angles to each other and forty-five degrees from the main line axis. These branches extended 283 feet further from shore, where the depth was approximately fifty-six feet. The branches had special fittings so that the sewage would discharge upward at a point about three feet above the ocean bottom.

By December 31, 1924, sewage flow to Hyperion was 110 MGD. The combined outfalls and screening plants were designed to last twenty-five years and, in the words of John Alden Griffin, "would serve a population exceeding two million people." The outfalls and screening plants did last twenty-five years, but long before that—and long before the population grew to two million—the sewage plant was overburdened.

In the summer of 1923, before construction of the screening plant but after passage of the bond measure to fund the new improvements, a Metropolitan Sanitation District for Los Angeles County was created, following passage of state legislation sponsored by Assemblyman Hugh Pomeroy, who represented a district in the Redondo-Hermosa Beach area. Representatives from the city and the county agreed to divide the county into two major divisions. All areas within the city, as well as those cities and unincorporated areas of the county north and west of Los Angeles, would be served by the city. The remainder of the county was combined under an umbrella organization, which became known as the Los Angeles County Sanitation Districts, an intergovernmental agency in which representatives from each district served on its governing board.

A joint administrative agreement permitted the Los Angeles County Sanitation Districts to be administered by a chief engineer located in Whittier. The South Bay Cities Sanitation District was the first to be formed under the new legislation. It included El Segundo, Palos Verdes Estates, Redondo, Hermosa, Manhattan Beach and a portion of Torrance. The new district quickly signed a contract with Los Angeles to carry its sewage to the new plant at Hyperion. Soon, other cities adjacent to Los Angeles lined up to connect their sewers to the Los Angeles system. Vernon, Culver City, Glendale, Venice, San Fernando, Beverly Hills, Santa Monica, Burbank and portions of Alhambra not served by the Pasadena sewage farm signed contracts with the city between 1922 and 1928.

Shortly after the creation of the Metropolitan Sanitation District, the county and city became engaged in a dispute about who would provide sewage treatment for Wilmington and San Pedro. A frequent critic of the city's sewerage efforts, County Engineer A.M. Rawn attributed the dispute to "an antipathy" which "engendered distrust, manifested, naturally by the weaker and less powerful." Noting the popularity and influence of then City Engineer Van Norman, Rawn wrote that Knowlton and H.G. Smith, a sewer design engineer in the city engineer's office, "held themselves somewhat aloof from active participation with the Districts. They would brook no criticism of the City's sewerage system despite the deplorable condition which existed . . . It seemed to disturb them to have the Districts embark upon a plan which had as major objectives the correction of errors which were so evident in the City's lines and outfall."

The disputes were settled in 1925, when the county submitted a plan to the California State Board of Health for upstream treatment plants and a joint outfall that would empty into the ocean at White Point off the south shore of the Palos Verdes Hills. The two municipal agencies agreed that San Pedro and Wilmington would be connected to the treatment plant at Terminal Island.

The new North Outfall included an East Los Angeles interceptor and a La Cienega Relief Sewer, reaching 17.5 miles. The North Outfall would eventually reach fifty-five miles to become the longest sewer outfall in the country. Composed primarily of semi-elliptical concrete, the sewer was lined with vitrified clay tiles.

On August 30, 1924, Van Norman was honored with a dinner party held in the far western section of the new North Outfall Sewer, perhaps the first and last such sewer party ever. Lights were strung along the length of a two-mile stretch and, according to reports of observers, "a strong draft, sweeping uphill from the sea, wafted away the cigar smoke and the fumes of the photographers' flashlights." By all accounts it was a festive affair, according to the report in the *Los Angeles Times.*

While Van Norman was completing work on the North Outfall, Griffin continued to have problems with the board. For several months in 1924 the City Council had requested that the City Efficiency Department conduct surveys of the city engineer's department. As in the case of City Engineer Hamlin, the surveys were extremely critical of Griffin and alleged that his office was disorganized and lacked supervision.

On October 10, the Board of Public Works unanimously terminated the service of the city engineer, effective October 16, 1924, and appointed H.A. Van Norman as city engineer and fixed his salary at twelve thousand dollars a year. Impressed with his work on the North Outfall, the board wanted Van Norman to establish a coherent and efficient engineering department. Griffin was notified of his firing by a reporter who had attended the Board of Public Works meeting where the surprise announcement had been made. He did not go gently into the night. He filed a letter with the Board on October 24.

"I was greatly surprised to receive the first inkling of dissatisfaction of my services," he wrote, "with its resultant removal from office by your Honorable Board, from a newspaper representative, and could hardly believe that you gentlemen, who have always treated me so friendly, would fail to at least show me the courtesy of informing me of your dissatisfaction and giving me the opportunity to resign, which I assure you I would have done on the spot." Griffin described his record of achievements while city engineer, taking credit for a variety of projects and concluded, with sorrow, that he had to "sacrifice the credit and glory for their accomplishment which rightfully belongs to me." Griffin's letter was received and filed.

Upon assuming office, Van Norman continued work on the outfall. The first fourteen sections—the main trunk line of the North Outfall sewer, extending from downtown Los Angeles to Hyperion—were completed by the end of 1924. The La Cienega Extension was completed a few months later in June. Additional extensions, including one to Van Nuys, were built within a few years. When completed, the mammoth sewer resembled a snake winding its way around the Santa Monica Mountains, slithering alongside the Los Angeles River to downtown, meandering southwest to the Playa del Rey hills, where it made its way beneath the hills shadowing the wetlands of Ballona Creek, then crossing under Pershing Drive to Vista del Mar Boulevard, a street paralleling the coast, before dying, disgorging itself into the ocean- bordered screening plant, Hyperion, where it joined the contents of the Central Outfall Sewer, emptying sewage into Santa Monica Bay.

Several months after completing the North Outfall, City Engineer Van Norman ran into trouble attempting to establish new, strict rules on work ethics and use of overtime. In an era when political patronage was on the rise, Van Norman's attempts to strengthen the integrity of the Engineering Department were not looked upon with the greatest of favor by council members. Reminding the Board of Public Works that he had originally accepted his appointment with the understanding that he would return to the Water Department to assist in construction of the new Colorado River Aqueduct, he served notice that he would resign after completion of the North Outfall. In August, with the blessings of the Board of Public Works, he returned to his job with the Water Department and John C. Shaw was appointed city engineer.

Shaw was appointed despite common belief among employees of the engineering department that Homer Smith, who had been a division engineer working under Van Norman, would be appointed to the position. Shaw was extremely well-connected to the political leaders of the city and had served on the Board of Public Utilities and Transportation before his appointment. Before joining the city he had been devoted to construction work, notably having nothing to do with sewers. Immediately before joining the engineering department, Shaw had been an assistant harbor engineer for less than six months.

Shortly after being connected to the screening plants, the submarine outfall began to fall apart. When the submarine outfall was laid in the ocean, the joints connecting the sections apparently broke in several places. Leaks from broken sections began appearing as a sewage field in the surf almost immediately. Sewage could be seen on the beaches as early as 1925.

The City took bacteriological tests and confirmed the presence of a sewage field with a count of more than "ten *B. coli* bacteria per cubic centimeter." The presence of *Balantidium coli* as a measure of contamination would later be replaced by the familiar enterobacterium *Escherichia coli (E. coli)* as an indicator of fecal pollution. According to the city engineer, this count was well within the California State Board of Health limits. *B. coli* is not common in humans, occurring most often in hogs, but its large size and frequent appearance in fecal matter made it easy to identify when bacteriological testing was in its infancy. There were no tests to identify the presence of *E. coli* bacteria, now used regularly as an indicator of fecal contamination. At the time the screening plants were designed, the engineers had planned to build an incinerator to burn the screened solids and debris taken from the effluent. In the interim, city workers buried the material in the sand dunes behind the screening plants.

To operate the screening plants and supervise the removal of the screenings, in 1924 the city hired Harvey Van Norman's nephew, James Van Norman, to work at the new screening plant. James was uneducated, and this job seemed an ideal match for him. He remained on the job for twenty-five years, until his position became obsolete after a new treatment plant was built. Describing his days at Hyperion in a 1969 interview, he said "the screens were hell to keep clean. The grease deposits would clog them up. Then they would go sour and they wouldn't do a damn thing. They'd just get all fuzzed up and we had to take them out of service and put in new ones." He explained that they took the screenings and "blew them by compressed air up to the burying grounds," which were directly adjacent to the screening plants. He added, "they were covered by five to ten feet of sand in the excavation that I dug with the clamshell bucket."

James Van Norman eventually became plant superintendent, and remained in that job until the construction of a new activated sludge wastewater treatment plant in 1950. The city built a house for him on the plant grounds and he lived there rent-free from 1930 to 1940, until the smells during the evening hours drove even James away from the plant.

After the screening plants were built, critics continued to castigate Los Angeles city officials for discarding a potentially valuable product while polluting recreational areas. In a 1926 article published in the Municipal League's *Bulletin*, Knowlton defended the operations of the screening plants at Hyperion. He argued that sewage odors had been eliminated, although he acknowledged that "at times there has been a sewage field at the outer end of the ocean section of the sewer." That sewage field continued to grow over the next several years and the slick, extending for more than a mile, was visible from the air as well as from the beach.

State authorities and the city continued to receive a barrage of complaints on a regular basis as the sewage field extended from beach to beach. Debris and scum were washing up on the beaches. On April 8, the city's health officer J.L. Pomeroy wrote that he would, "at once personally take up the matter of the uncovered-screenings nuisance in the sand hills at Hyperion." As a result, the city installed sewage grinders at the screening plant to decrease the amount of large uncovered screenings. The grinders did not eliminate either the debris or odors from the ever-expanding river of sewage pouring into Hyperion at the beach.

As complaints continued to pour in during 1928, City Engineer Shaw alleged that submarine oil springs several miles offshore were the source of the pollution. He and other city officials went by boat to examine a large oil slick several hundred yards west of the Chevron Refinery, which Shaw blamed for the ocean contamination. A few months after the initial investigation of the contamination along the beach, Shaw agreed to dispatch city workers to remove any traces of grease which found its way into the ocean.

In the years from 1925 to 1929, before the stock market crash, thousands of miles of sewer work were completed within the city. Sewage woes, however, did not end with the closing of the decade. More than fifty tons of wet screenings a day were being buried in the sand dunes behind the plant. The city began using an acid treatment to kill the incessant nuisance of the flies attracted to the smell of the screenings. Unfortunately, the treatment was intermittent and somewhat costly. Hyperion Plant Superintendent James Van Norman later recalled that the county sheriff told him to stop polluting the beach. "I thought that was pretty funny. What did they expect me to do?" he said, "I couldn't stop the flow of sewage coming into the plant."

GALLERY

Early Sewerage
1887-1945

City Engineer and Mayor Fred Eaton designed the city's first sewer system.

Sewer construction in the 1920s.
Top: Installing sewer in the 1920s.

Concrete pipe is ready to be installed. 1925

Top: Tunnel work proceeds on 1924 North Outfall Sewer.

Framework for a sewage pumping plant in Mar Vista. circa 1927

Finished Mar Vista pumping plant with landscaping.

Brick lining of old city sewer.

Tile work at the 1925 Hyperion Screening Plant.

Top: Construction of Hyperion Screening Plant. The Pacific Ocean, contaminated with sewage, is in the background. 1924

Middle: South Screening Plant at Hyperion. circa 1925

Cleaning sewage screens in 1930s.

Top left: A concrete sewer section.

Top right: City Engineer Lloyd Aldrich was appointed in 1933.

Raw sewage is diverted into Santa Monica Bay in 1937.

Top: Sewage pipe is laid in the harbor at Terminal Island. 1933

Workers prepare to ride a boat through the North Outfall Sewer. circa 1938

A food stand was adjacent to the Sewage Screening Plant at Hyperion.

Top: In the Depression era, people fished off the sewage pier adjacent to the sewage screening plant at Hyperion.

A 1937 gas survey truck.

Top: An inspector sniffs for gas from a sewage manhole.

The Works Progress Administration put citizens to work on the sewers during the
Great Depression.

BROWN ACRES

Floods in 1938 popped manhole covers as sewage spread through the streets.

CHAPTER EIGHT

Pollution of Santa Monica Bay

The effervescent bubble that was Los Angeles in the 1920s burst with the arrival of the Great Depression following the stock-market crash of 1929. In the decade of prosperity, Mayor George Cryer, a bland but good-looking man, had served three terms. During that time he presided over widespread political corruption and untrammeled vice and crime. A reform candidate, John C. Porter, a used-car dealer with no political experience, was elected mayor in December 1929.

In the tradition of city politics, shortly after taking office, Mayor Porter fired City Engineer John Shaw. Claiming that Shaw was a structural engineer at a time when sewage expertise was needed, Porter hired John Jessup as a man "who could cut the red tape and get more work done in the department." Jessup, who was sixty-five at the time, had previously served as city engineer of Berkeley and Whittier and had most recently maintained a private practice, specializing in sewage and garbage handling. He was also a close friend of Board of Public Works Commissioner William Hyde.

Within a few months, Jessup presented a plan to rectify the ongoing pollution at Hyperion. He instituted minor repairs of the submarine outfall, but the results were minimal. The disgusting condition at Hyperion continued unabated. As the effects of the Depression hit even somewhat isolated Los Angeles, in October 1930, voters approved a general relief bond of six million dollars, half of which went to the Bureau of Engineering to hire day laborers for work projects designed by the department. Although the city launched a sewer-bond campaign for six million dollars to fund improvements to the entire system, the measure failed to get the necessary two-thirds majority. With the relief bond money, work was continued on extension of the North Outfall Sewer despite the failure of the bond measure.

Another bond measure to raise the six million dollars was proposed in June 1931. The city solicited and received the support of dozens of community and civic organizations. With the approval and support of these organizations and all the city's newspapers, the bond measure received a majority of votes, but again failed to get the necessary two-thirds. These were hard times, and an increase in taxes for sewage improvements was a low priority. To meet the unemployment crisis, an additional relief program was started with approval of a five-million-dollar bond measure on the same ballot as the sewage measure.

In 1931 the city added five additional drum screens at the north plant and built an experimental skimming tank to reduce the amount of grease clogging the screens. Discharge of industrial wastes from meat-packing companies was prohibited by city ordinance, but fruit and vegetable canneries continued to discharge solid waste into the sewers, creating even greater pressure on the overloaded system.

Also in 1931, sanitary engineer Ray Goudey embarked on a mission to reclaim sewage water, a first for the city. He was a well-known and widely respected engineer who had previously worked for the California State Board of Health and was now working for the city's Department of Water and Power (DWP). In 1931, he installed a water reclamation plant in Griffith Park that treated approximately two hundred thousand gallons per day. The plant used sewage from the city's North Outfall Sewer.

Goudey removed the sludge from the water by adding coagulants to the incoming sewage and then sent the water to settling tanks, where the sludge was removed. The extricated sludge was placed in anaerobic tanks to digest the volume of the material. The gas produced in the anaerobic tanks was released to the air, while the sludge was offered as fertilizer to the public. The sewage water was then filtered, chlorinated and finally dechlorinated before it was released into the Los Angeles River.

Several members of the Los Angeles Water Department visited Goudey's plant in September 1931, and each drank a sample of the reclaimed water for the benefit of the press contingent. Apparently no one got sick. The project raised hopes in the city engineer's office that finally the city might reclaim some of the sewage, rather than just pour it into the ocean.

In a 1933 report to the Board of Public Works, Knowlton, the Bureau of Engineering's sewage engineer, outlined a plan to divert 25 MGD from the North Outfall in an area of the San Fernando Valley for irrigation or for replenishment of groundwater supplies. He subsequently solicited unofficial bids for work and estimates for construction of lines for reclaimed water. Unfortunately, his report died when city officials looked elsewhere for a supply of water. Even more unfortunately, it would take more than thirty years for Los Angeles to build water reclamation plants.

Despite the enthusiasm of the press and the unqualified support of county and city engineers for continuing Goudey's program, the plant in Griffith Park was closed within a few years with little public notice. That may have been because the Metropolitan Water District was then seeking water from the Colorado River to supply the unquenchable thirst of Los Angeles. In the years before the aqueduct was completed in 1941, the district was uninterested in reclaiming the wastewater. Ironically, although no one discussed the fact, water from thousands of cesspools and inland treatment plants was percolating into the water table, providing a rough kind of water reclamation.

By 1932, the overflows at Hyperion had grown so serious that the city built a new bypass channel to Ballona Creek. Leaks from broken sections of the underwater outfall continued to wash back onshore. Overflowing sewers in several parts of the city backed up sewage into people's homes, just as it did before construction of the North Outfall. Promises were made to stop the outfall leaks as complaints continued unabated from the mayors of El Segundo, Redondo Beach, Manhattan Beach and Hermosa Beach. The promises were never met.

A groundswell of opposition to Mayor Porter's self-righteous politics developed during the first years of the Depression. At the time of the 1933 mayoral campaign, people had come to mistrust Porter's optimism, his coterie of influential lawyers and businessmen who were running the city, and also his plan for privatization of public utilities. Frank Shaw, Chairman of the County Board of Supervisors, was riding a wave of popularity, due in large part to his support for county-wide relief for the unemployed. He was seen as a fresh breeze, unconnected to either the Porter coalition or the *Los Angeles Times*. Shaw was elected in June 1933 in a close contest.

Besides the replacement of many of the city's commissioners, newly elected Mayor Shaw replaced the sixty-eight- year-old City Engineer Jessup with an energetic and aggressive civil engineer, Lloyd Aldrich, who was then forty- seven years old and already had twenty-four years experience.

Aldrich took office in August 1933, brimming with confidence and the can-do attitude that was the unwritten requirement to face the challenges of multiple public works projects, an ailing sewer system and the need to put thousands of unemployed men to work. Like Fred Eaton, Aldrich was a visionary. He was also an excellent administrator and superb politician. Unlike many of his predecessors, he was able to create wide support for his sewage programs and rebuff efforts to reduce or modify his plans to create a new sewage treatment plant.

Aldrich reorganized the department, began planning for a new wastewater treatment plant at Hyperion and within the year began a study of the condition of the North Outfall Sewer. Under the direction of President Franklin D. Roosevelt, federal money from an alphabet soup of agencies—the CWA (Civil Works Administration), PWA (Public Works Administration), WPA (Works Progress Administration) and State Highway and State Emergency Relief Agencies—poured into the city. Lloyd Aldrich was in charge of millions of dollars that would support thousands of workers on hundreds of public-works construction projects. Sewers, bridges, streets, a waste-activated treatment facility at Terminal Island Treatment Plant, libraries, fire stations, tree planting and writers' projects were completed during these years.

During the 1930s, the mayors of El Segundo, Manhattan Beach, Hermosa Beach and Torrance demanded, to no avail, that actions be taken to clean up the beaches. Sewer workers at Hyperion began burning the screenings from the plant in an incinerator early in 1933. After a major storm flooded the screening plant in January, a bypass around the plant was constructed so that in case of heavy storm water incursion all sewage could be routed directly to the ocean.

In August, Ballona Creek once again became the outlet of choice for the overburdened North Outfall. Aldrich recommended building a gate chamber in the North Outfall to allow for overflow to go to Ballona Creek after chlorination. Despite protests from the county health department, the gate was built. The health department believed that, even with chlorination, large amounts of sewage flowing into Ballona Creek would not be healthy. They were right. The gate would be used on many occasions in the future and city engineers would often release millions of gallons of raw sewage into the creek, frequently without the mitigating use of chlorine. The gate at Ballona Creek was an easy but unfortunate solution that led to decades of pollution.

In 1935 Aldrich requested federal money from the Works Progress Administration (WPA) to construct a modern sewage plant. The request for $7,750,000 covered installation of sedimentation tanks, skimming tanks and venturi flumes. His plan called for sludge to be pumped to digesters before going to vacuum filters for drying. The dried sludge would then be burned, using sludge gas gathered from the digesters. The WPA granted his request (shy three-quarters of a million dollars) and appropriated seven million dollars for construction of the new wastewater treatment plant.

Although plans for the replacement plant were extensively described in several articles published in 1936 and 1937, it was never built. Instead, four million cubic yards of sand, including the screenings buried prior to the installation of the incinerator, were removed by several hundred men who were employed to dig it out by hand. The Hyperion site was uneven and sand dunes surrounded the old screening plant. So, these sand dunes had to be excavated before any construction could begin. The screenings were then barged out to sea. The removed sand was used for expansion of the beach adjacent to the plant.

Eventually, sometime in 1938, an experimental treatment plant was built on the excavated area. This plant diverted and treated twenty thousand gallons of the more than 110 MGD of sewage rushing into Hyperion, one-tenth of what the Water Department's engineer Goudey had been reclaiming in 1931. Although Aldrich claimed that the experimental plant at Hyperion provided valuable information for the design of a larger plant that would treat all the city's sewage, seven million dollars was a lot of money for very little.

There was never any suggestion that Aldrich had misused the money, but he was rapidly gaining a reputation for a close relationship with construction firms in the city, and it seems likely that the plant was something of a boondoggle for a local contractor. Aldrich's annual reports did not clarify how or why the experimental plant cost so much to build.

In 1936, Reuben Brown, an assistant superintendent of sewer maintenance, went on an unprecedented trip through six miles of the North Outfall, north of the Hyperion plant. He was lowered in a specially designed boat equipped with removable air tanks attached to the side of the boat. He floated through the concrete sewer for several miles during low flow. His instruments included a radiotelephone, an oxygen helmet, a searchlight and a camera. Forced ventilation was provided with the aid of portable blowers. It was a unique procedure. Later, sewer workers often used a smaller boat while wearing gas masks as they rode through the sewer examining the walls. The images of the heroic engineer "who braved death and defied the hydrogen sulfide gas in the tunnel," as local newspapers described him, captured the public imagination for several days in July. He made the trip in a series of descents into the outfall and took many photographs and notes about the condition of the sewer.

Brown found extensive damage: cement joints were soft and a longitudinal crack about 480 feet long was found about 1400 feet before the screening plant. In other areas, sulfuric acid attacked the Portland cement mortar joints between the clay-tile blocks, allowing extensive spalling. Spalling is a term used to describe breakage of individual tiles. Once the tile was completely spalled, hydrochloric acid in the sewer attacked the exposed surfaces and unreinforced concrete underneath the tiles began to crumble. Reinforced-concrete repairs were not made until well into 1942. The 1-Mile submarine outfall, lying in sixty-five feet of water, was extensively repaired during 1936. Additional concrete supports were installed to raise the bottom of the pipe and a temporary pier was built to carry the pipe.

The storm drains had been separate from the sewer system ever since Fred Eaton prevailed in his determination to design a system that separated storm runoff from the sewage. In March 1938 a major storm predictably overwhelmed most of the southwestern parts of the city, causing the familiar Yellowstone-like eruptions from manholes, as streets were once again flooded with the combination of storm water and sewage. Several major sewer lines, including a large interceptor sewer in the San Fernando Valley, were washed out for several miles and a sewage pumping plant was completely washed away. The temporary pier over the submarine outfall in the ocean was also washed out. In many areas of the city, pipelines and structures were damaged where slides occurred. A second pier was constructed and the outfall was suspended from the new pier.

In temperate weather, people came to the new pier and purchased fishing tackle and even bought snacks from a nearby beach refreshment stand. Children would often play in the sewage-contaminated waters not far from the sewage pier. Beachgoers avoided the stretch of sand between Ballona Creek and El Segundo, but continued to use sewage-contaminated beaches as far south as Venice, which by then were dominated by hundreds of oil wells creating a massive black spiderweb of derricks adjacent to the beach.

Meanwhile, corruption, vice and criminal activities were creating a dilemma for the Midwestern, mostly law-abiding community of Los Angeles. Reformers organized once again, this time forming a coalition of forces that included Clifford Clinton, (the wealthy owner of a chain of popular cafeterias, who formed a citizens' investigating committee) John Anson Ford (a county supervisor and leader of anti-corruption forces in the city) and a variety of liberals and labor leaders.

In 1938, a series of scandals rocked the city. Two police officers and the Chief of the Police Intelligence Bureau were convicted of the car-bombing of an investigator hired by Clinton. The general manager of the Civil Service Commission was suspended for altering exam papers and six firemen were fired for taking payoffs. Even Aldrich was attacked, ironically for allegedly forcing some of his employees to circulate petitions supporting his earlier successful drive for civil service protection.

The coalition turned to respected judge Fletcher Bowron to oppose Shaw in a recall election in September 1938. Shaw was recalled, and Bowron was elected by an overwhelming two-to-one vote. Bowron, who called himself a New Deal Republican, appointed a new police commission. Shortly thereafter the police chief resigned. Unfortunately, in the same election, a sewage tax measure to construct the North Central Outfall and a new sewage treatment plant at Hyperion failed by a vote of 202,899 to 102,280.

Under Aldrich's supervision the department of engineering was able

to make some improvements to the sewer system. During 1938, illegal connections from roofs, yards, washracks and the like were eliminated, and manhole
covers adjacent to gutters were sealed, in a less-than-successful effort to stem
the incursion of the storm water. To exhaust air from the sewer and to slow the
continuing deterioration of the outfall along the six-mile section, a fifty-four-
inch fan powered by a seventy-five-horsepower diesel engine was installed on
the North Outfall Sewer about one thousand feet west of Centinela Boulevard.
Several hundred tests for explosive sewer gas were made daily. Occasionally,
inspectors would lean over manholes and sniff for gas. No explosive conditions
were found, but many a motorist must have been startled by the sight of a
sewage-sniffing man huddled over a manhole.

By the end of the year screenings from the plant were no longer buried
in the dunes. Instead, they were sold to a contractor, who dried the pressed
screenings with a direct-heat rotary drier so the finished product could be used
for fertilizer in the citrus groves of the San Fernando Valley.

In 1939 the Board of Public Works hired a special Board of Engineers
to examine Aldrich's plans for treatment of the sewage. The three eminent engineers were Samuel Greeley, Charles Hyde and Franklin Thomas, then dean
of the California Institute of Technology. The board agreed with City Engineer
Aldrich that the city should build a new treatment plant at Hyperion, build a
new overflow structure for the North Outfall at Ballona Creek and repair both
the Central and the North Outfall. Nevertheless, it also strongly urged the city
to immediately purchase three additional sites for new treatment plants. The
board suggested that one plant should be built in the San Fernando Valley, one
near the Los Angeles River east of downtown Los Angeles and one in the general area of the Baldwin Hills. It was lukewarm about Aldrich's desire to build
a North Central Outfall to relieve the burden on the Central Outfall.

After Shaw's recall from office, Aldrich was subject to a barrage of attacks from Bowron's Board of Public Works. By way of a charter amendment
Aldrich had received permanent protection of civil service status in 1937. Reuben Borough was appointed to the board in 1939 and was a frequent opponent
of the city's sewage plans. First, Borough attempted to end Aldrich's practice of
awarding lucrative street-paving contracts without benefit of competitive bid-
ding. He was unable to break the paving trust, but was successful in removing
a corrupt street superintendent. He was also mistrustful of Aldrich's plans for

a new treatment plant.

Despite the attacks, Aldrich was Teflon. He was never personally accused of corruption and responded to every attack with a dash of withering scorn. After surviving these attacks, Aldrich decided to go ahead with his design for a new treatment plant, using the Board of Engineers report only when it suited his own plans. He ignored the suggestion to build upstream plants in favor of a new, state-of-the-art activated waste treatment plant.

The California State Board of Health granted the city a permit on September 3, 1940 to build a "temporary" two-year bypass of 100 MGD at Jackson Avenue where it met Ballona Creek. It would continue to be used as an escape valve well into the late 1980s. At the same time, the board ordered the city to finance a program of construction and install remedial works at Hyperion. It suspended the 1923 permit to discharge the screened sewage at Hyperion. Unfortunately, the Board could condemn, but not enforce. Despite the suspended permit, the sewage continued to flow unabated into the ocean and surrounding beaches.

Meanwhile, the feud between Aldrich and the Bowron-appointed Board of Public Works continued unabated. In January 1941, the board attempted to diminish the scope of Aldrich's Department of Engineering, which was now called the Bureau of Engineering. They created two new bureaus, the Bureau of Street Maintenance and the Bureau of Sanitation. The latter was responsible for operating and maintaining the sewer system. Although this was an effort to diminish Aldrich's power, he continued to hold sway over the activities of the newly formed Bureau of Sanitation, headed by former Bureau of Engineering employee Warren Schneider.

The Los Angeles Times supported Aldrich and censured the Board of Public Works in a lengthy jeremiad that prophesied doom and gloom stemming from the board's actions. The board then created a new Bureau of Accounting, which the newspaper wrote would cause "Aldrich, denied of his intelligence service, to commit some blunder that can be used to bring about his dismissal." In fact, Aldrich's accounting was impossible to decipher. Every year he created lists of the number of employees working on different projects. The numbers would be allocated to specific projects, but his staff created no balance sheets. Even the simplest rules of bookkeeping were ignored in favor of these multi- page lists that were clearly an attempt to obfuscate, not illuminate.

There has never been any evidence that Aldrich or bureau employees were profiting from the confusing bookkeeping. It just seemed another way for the city engineer to control expenditures of the department. At the same time he greatly expanded the bureau's annual reports, which were filled with flowery language extolling work under Aldrich's supervision.

In 1941, in an effort to halt the Board's attacks on Aldrich, his allies on the City Council placed a charter amendment on the April ballot that would eliminate the Board of Public Works and put its functions under the control of the city engineer. The council also approved a bond measure to assess for twelve years an annual tax levy of nine cents per hundred dollars of assessed valuation, to provide for the sewage improvements recommended by the Board of Engineers. Borough had a weekly radio program, during which he usually discussed public-works projects and often attacked Aldrich. Despite his enmity toward the city engineer, Borough supported the tax levy, saying, "the plan will transform our present menacing 130 million gallons per day of raw sewage poured into the ocean into a community asset." Borough argued that the plan would be a pay-as-you-go plan.

Despite the support of the Board of Public Works and local newspapers, voters rejected the seventeen-million-dollar measure that would have been used for a wide variety of sewage projects, including improvements at the Hyperion site. In the primary election, the measure to eliminate the Board of Public Works was also defeated. Mayor Bowron, however, was re-elected.

In our current era of eco-politics and environmental awareness it may seem extraordinary, but in 1941 the general public treated pollution of the beaches with surprising indifference. Most would simply avoid the black tar which had washed onto the shore from ocean oil wells. Although residents of the beach communities complained, tourists and the citizens of Los Angeles flocked to beaches from Venice to Hermosa Beach, polluted north and south of the screening plant by raw sewage.

When Japan bombed Pearl Harbor on December 7, everyone knew there would be no federal sewage funds until the war was over. During World War II, however, complaints from adjacent beach communities continued to pour into the California State Board of Health. The Board began an extensive study of pollution in Santa Monica Bay in 1942. Shortly after the study began, Elmer Belt, M.D., who was the director and president of the California State

Board of Health, warned that it was only a matter of time before the beaches were quarantined. In May the board revoked the city's permit to operate the Hyperion plant. This was a quixotic gesture at best. In a May 1942 report, portions of which were quoted in the *Los Angeles Times*, Dr. Belt wrote,

> There is visible evidence of solid fecal material rolling in the water. It is terrific and slimy off Hyperion . . . We get constant complaints from mothers about children who have visited the beach then have been unable to attend school for three weeks because of dysentery. Yet no definite epidemics have been directly traced to the sewage. No typhoid or other plague has resulted. Visitors carefully avoid the black tar and unwittingly walk and sit on waves of sand amid rows of sewage grease. Later they are puzzled as to how they became so dirty and greasy.

That same month, Aldrich went to Washington in an effort to obtain eight million dollars to repair the North Outfall, build a new North Central Outfall, repair other lines and build a new treatment plant. The federal government was directing all its attention to the war effort and in June the Federal Works Agency turned down the city's request for funds.

By September 1942, sewage flowing into Hyperion had reached 141 MGD. On April 3, 1943, the California State Board of Health quarantined Southern California beaches for ten miles along Santa Monica Bay from Fourteenth Street in Hermosa Beach to Brooks Avenue, north of the Venice pier in Los Angeles. Signs were posted every few hundred feet to warn beachgoers. Later, they would extend the quarantine two additional miles. The quarantine lasted eight years.

On June 26, 1943, the California State Board of Health released a seventy-page *j'accuse*. The "Report on A Pollution Survey of Santa Monica Beaches," presented the results of a detailed investigation of the city's sewage disaster. Documented with tables, lists of stations surveyed, bacteriological analyses, as well as photos of children playing in the sewage-laden sand, the report was a call to arms. Excoriating the city for the fetid beaches, the Board noted that the North Outfall "is cracked to the point of bursting," and that the population now reached 1,787,000, making the "fine screen method of sewage treatment totally inadequate" Beachgoers were threatened with arrest if they ignored the quarantine. Although they did avoid the area directly opposite the Hyperion plant, people continued to visit beaches from Venice to Hermosa Beach. Fortunately, as Dr. Belt reported, the bacteria in the sewage-laden ocean were not typhoid or cholera bacilli.

Following the release of the board's report, Aldrich Blake and Ernest Chamberlain, campaign workers for Clifford Clinton, produced a small pamphlet, *Sewers of Los Angeles*, subtitled, "The Story of Millions in Damages, Disease, Pestilence, Death." The pamphlet sold for ten cents. The authors did not mince words. "Welcome tourists!" they wrote, "Thine eyes shall behold the greatest outdoor privy in the world—a vast reservoir from which may emerge the three most deadly diseases of man—hookworm, typhus and cholera." With public apathy about the sewage menace at a persistent level, it seems unlikely that the pamphlet found much of an audience.

In August 1943, Aldrich Blake sued to force the city to stop using the Hyperion Screening Plant. He asked the court to arrest the city engineer and members of the City Council. On December 18, Municipal Court Judge Charles B. MacCoy issued an interim opinion on Blake's suit, in which he suggested that the city should join the County Sanitation Districts. Although Judge MacCoy took special note of City Engineer Aldrich's lack of cooperation, he did not issue any warrant or impose any fines. Three years later, the judge, a resident of Playa del Rey, would himself file a nuisance suit against city officials in an effort to forestall construction of the Hyperion plant. His efforts, like those of Blake, were ineffective.

In October 1943, the California State Board of Health rescinded the permits of nine of the tributary cities to use the Hyperion outfall sewer. Nevertheless, sewage continued to flow to the plant.

CHAPTER NINE

A Modern
Treatment Plant

As the California State Board of Health attempted to force city officials to do something about the Hyperion mess, Lloyd Aldrich and Mayor Bowron continued their battle. Initially, Aldrich proposed that the council adopt a ten-million-dollar bond measure for a variety of sewage projects, of which $8,566,000 would be used to build a new treatment plant. The City Council Public Works Committee killed the bond measure. In May 1943, Aldrich asked the council to approve a sewer service charge of one million dollars to repair the ailing sewer system. The sewer service charge was adopted by the City Council, but was vetoed by Bowron, who insisted that Aldrich had not consulted him about his plans for sewer improvements. This was undoubtedly true, since Aldrich would frequently show public contempt for the mayor who had replaced Aldrich's friend, Frank Shaw.

In December, upon the recommendation of newly appointed Board of
Public Works member Frank Gillelen, himself a civil engineer, Mayor Bowron
and the City Council agreed to hire the renowned firm of Metcalf & Eddy to
study the problem. Metcalf & Eddy had extensive experience in sewering services for major cities throughout the United States, and its namesake partners
were the authors of an authoritative three-volume textbook on sewerage standards and practices.

Metcalf & Eddy presented its report to the Board of Public Works in
April 1944. Although the company supported Aldrich's plan for a high-rate
activated sludge treatment plant, they disagreed with his desire to put the plant
close to the beach. Instead, they recommended that the city buy two hundred
acres east of the sand dunes and site the new plant there, leaving the existing
site for development and use as a beach. Moreover, as the engineering board of
Greeley, Hyde and Thomas had suggested in 1939, the firm strongly recommended that the city build an upstream plant in the industrial area near the Los
Angeles River. Its opinion was that the flow to Hyperion would be greater than
the approximately 240 MGD Aldrich was predicting. Planning experts believed
that the population would reach three million by the mid-1960s.

Aldrich had estimated that a new plant and repairs to the North Outfall
would cost about eight million dollars. Metcalf & Eddy estimated that improving
the sewer system, building the new plant at Hyperion and buying land for the
second plant would cost approximately twenty-one million dollars. The City
Council and Board of Public Works adopted its report the following May.

Only a few weeks before Metcalf & Eddy presented its report, the state
attorney general sued the city to enforce the California State Board of Health
restrictions. In January 1946, Judge Vickers ruled in favor of the state and
ordered the city to build a new submarine outfall and a complete sewage treatment plant. The judgment was moot, since the city was moving forward on
plans to fix the sewage mess. In addition, he ordered the cities and agencies
connected to Los Angeles' sewer system to pay a proportionate share of the
new costs and future maintenance. The city became embroiled in decades of
futile efforts to get the agencies to pay their fair share, before finally succeeding
in the late 1980s.

Following Judge Vickers' ruling, the South Bay Cities District disconnected from the city system and in 1950 connected to the Los Angeles County

system. Seventeen other cities and unincorporated areas of the county remained connected to the city system.

On April 3, 1945, as World War II drew to a close in Europe, Los Angeles voters finally approved a ten-million-dollar bond measure to build a new treatment plant, acquire any necessary additional land and construct a new submarine outfall at Hyperion. Aldrich wanted to build the new plant close to the beach. The City Planning Commission went against Aldrich and recommended that the easterly site endorsed by Metcalf & Eddy should be used, in hope of creating "a large seashore recreational area, similar to Jones Beach in New York in which band concerts and beautiful scenery might be provided for thousands of people."

By this time Aldrich was an extremely influential and powerful political figure who could count on the support of the City Council, which often warred with Mayor Bowron. He ignored both the Planning Commission and Metcalf & Eddy's wishes regarding siting of the new plant. Instead he insisted on building the new treatment plant on the same site as the old screening plant. Metcalf & Eddy disputed the city engineer's argument that moving the plant would cost an additional million dollars. Nevertheless, the Board of Public Works approved Aldrich's plans.

In the general municipal election on May 27, 1946, voters unanimously approved an additional ten million dollars for sewage improvements to ensure completion of the treatment plant and to provide for additional relief interceptor sewers. On the same day, a contract for $3,517,000 went to Aldrich's friend Guy F. Atkinson, low bidder for construction of the new twelve-foot-diameter submarine outfall, to extend one mile from the shoreline. A second contract, for $3,494,088, was awarded a few days later to Kiewit & Sons Construction to excavate fourteen thousand cubic yards of sand at the site. The earlier work by hundreds of laborers from 1935-1938 was apparently insufficient. No other sewage project ever entailed so much moving of sand.

Several months before the contracts were awarded, Gillelen, now president of the Board of Public Works and apparently worried that Aldrich was not following the recommendations of Metcalf & Eddy, appointed a board of consulting engineers to evaluate Aldrich's specifications and plans. The consultants included the professor and dean of students at California Institute of Technology, Franklin Thomas, Charles T. Leeds of the engineering firm of Leeds and

Hill, and Los Angeles County Chief Engineer A.M. Rawn. Although the low bid for the new ocean outfall had been received in January, the award had been held up, pending the results from the consultants. Citing a condition of extreme emergency and noting that the low bid would have expired in May, Aldrich convinced the Board of Public Works to award the outfall contract to Atkinson.

Two months after the contract was awarded, the consultants presented their overwhelmingly critical report. Annoyed that the contract for the outfall had been awarded while their study was ongoing, they expressed additional concerns, many of which would later prove to be accurate. In an interview in 1966, Rawn recalled his dispute with Aldrich. Citing the county's two-hundred-foot-deep outfall off White Point, where the county disposed of primary-treated sewage, Rawn said, "We recommended that an outfall be built two and one-half or three miles out to sea instead of a five-thousand-foot outfall which was proposed. We felt that if this were done, there would be no beach contamination; the process would be far more satisfactory and more certain to achieve the results and far less expensive, both in construction and in operation."

The consultants concurred with Rawn and recommended that the outfall be extended two miles, instead of the proposed one mile, and suggested that the city provide only primary treatment to the sewage, which would have been far less expensive than the proposed plant. They predicted that "by 1957, the plant will have to operate only with partial secondary mixed with primary," due to an anticipated population increase. They were extraordinarily prescient. In the mid-1950s, the high-rate activated sludge plant was dramatically scaled down when two new outfalls, one for sludge and a five-mile outfall for reduced-treated effluent, were constructed. The plant was converted to seventy-five percent primary treatment with only twenty-five percent of the sewage receiving secondary, or activated sludge, treatment.

Aldrich ignored the consultants' recommendations. He was the strongest and most opinionated city engineer in the city's history. While others such as Eaton and Griffin were equally strong-minded, only Aldrich was successful in gaining the council's support for his plans for the city's infrastructure. Although many of the city's current engineers consider him to have been one of the best and brightest of their profession, Aldrich was essentially a canny politician and administrator who would follow his own path no matter what others suggested. In fact, his main legacy would be a multitude of public works projects completed with federal funds and a plant that was outdated within two years of its completion.

Nevertheless, the new plant would be one of the largest and most advanced treatment plants in the world. During construction in 1950, engineering and construction journals praised the project as the "World's Finest Sewage Treatment Plant." Up to 245 MGD of sewage would flow into the plant from the city's two main arteries.

The new plant was designed to remove all impurities from the water. The sewage would flow from the North Outfall and the Central Outfall to buildings where automatic bar screens would remove rags and wood. The material would go to a "launders" building where it would be finely ground and returned to the main flow of the sewage. (Launders are troughs in the secondary clarifiers through which the clean water is discharged as the sludge settles to the bottom.) The main flow would then go to a building that housed grit removal tanks, where sand and inorganic material would be removed and sent to a landfill.

The sewage would then be chlorinated before it would go to eight reinforced concrete primary tanks, each three hundred feet long and fifty-six and one-half feet wide, where sewage would settle for approximately two hours. There, an endless-chain pulley of wood planks would skim grease and oil off the tanks and remove settled solids from the bottom.

Most of the solids, now called sludge, would then be pumped to eighteen round, thirty-foot-high, closed anaerobic digester tanks. The sludge would be heated and would then go to wash tanks, where it would be washed to reduce its alkalinity. The sludge would then be filtered and dried so it could be packaged and sold as fertilizer. Vapors and gases would be vented through a two-hundred-foot stack.

The sewage water would then go to a series of thirty-two open aeration tanks, each three hundred feet long and thirty and one-half feet wide, into which air would be pumped. As an initial step, some sludge from the primary tanks would be added to these secondary aeration tanks. The air would allow bacteria in the water to feed on suspended solids that had not settled out in the primary tanks. Aeration would take approximately three and one-half hours. The aerated water would then go to twenty final clarification tanks, each 125 feet long and seventy-six feet wide. There, the engorged bacteria (called "activated sludge") produced in the aeration tank would settle and be removed. Some would be returned to the aeration system; the rest would be pumped back into the digesters.

By the end of June 1949, all basic design work on the new plant was finished. According to Aldrich, it was designed to serve the needs of three million people, based on an average dry-weather sewage flow of 245 MGD. He estimated that the total construction cost, including a mile-long submarine outfall, would amount to forty-one million dollars—very close to the final amount spent. Hailed as an outstanding example of cooperation between private contractors and public authorities, the plant was completed under Aldrich's direction by outside contractors who did the actual work on all the facilities. That year, Reuben Borough, who finally saw his dream of a waste activated sewage plant becoming a reality, said that his ten years service was enough and resigned from the Board of Public Works.

While construction of the new plant continued, Aldrich decided to challenge his critic and constant foe, Mayor Fletcher Bowron, who was now running for his third term. With the backing of several developers, Aldrich filed candidacy papers in March 1949. Aldrich, whose name was at the top of the April primary ballot, captured 87,957 votes to Bowron's 180,000. He forced Bowron into a runoff election because the combined vote for his ten challengers exceeded fifty percent. In what became an ad hominem exchange, the candidates took to the airwaves. Noting that the campaign would be decided on May 31, the *Los Angeles Times* described it as "one of the bitterest fights in the history of local politics."

Aldrich claimed that the city needed "constructive leadership and a change." He highlighted brutality by the police department against minorities and organized labor. He was particularly incensed by Bowron's efforts to claim credit for municipal improvements, including the construction of the Hyperion Treatment Plant. Bowron retorted that the city was "the cleanest city in America," while alleging that the campaign against him was carried out by "hoodlums, racketeers, moochers and broken-down political hacks." Bowron was able to garner support from most of the city's newspapers, as well as most of the middle-class districts of the city, while Aldrich was able to muster a strong base of support in working-class and minority neighborhoods.

Shortly before the election, Harvey Van Norman, former city engineer and general manager of the Department of Water and Power, announced his support for Bowron. The *Daily Mirror*, the *Los Angeles Times*, the *Daily News* and a semiweekly, the *Los Angeles Independent*, endorsed the mayor. Only the AFL labor newspaper remained loyal to Aldrich.

A few days before the general election, newspapers carried critical sto-ries about the new mile-long, twelve-foot- diameter ocean outfall on their front pages. The *Times* headlined its report on the new ocean outfall, "Hyperion Fiasco Laid to Aldrich." Tests on the new submarine outfall produced evidence of multiple leaks. "Faces $4,000,000 Loss in Sewer Project," proclaimed the front-page headline of the Hollywood edition of the *Los Angeles Independent*— along with a photograph of a heaving mass of raw sewage pouring from a hole in the old outfall about three hundred yards from the oceanfront. A geyser farther out along the outfall which spewed sewage two hundred feet into the air was shown in other newspapers. Despite this bad publicity, Mayor Bowron defeated challenger Aldrich by only thirty-two thousand votes in the general election in May.

After the election, newspaper stories continued to herald the difficul-ties that Guy Atkinson, builder of the submarine outfall, was having. Despite his friendship with Aldrich, Atkinson responded to charges by telling the *Los Angeles Times*, "You can't hire me to build a bucket with a hole in it and expect it to hold water," in an effort to place the blame for the leaky outfall on the city engineer's specifications. Apparently, oakum (a tar-impregnated loose fiber ob-tained by untwisting and picking apart old ropes) was initially used to seal the joints; when that failed, rubber was used. By June 30, leaks continued at a rate of twelve hundred gallons a minute.

Despite repeated concerns about leaks in the Atkinson-built subma-rine outfall and a year's delay in opening the pipeline, in November 1949 the city commissioned and accepted the new outfall when leaks were reduced to less than eight hundred gallons a minute. As primary treatment was initiated in May 1950, Mayor Bowron christened the forty- one-million-dollar plant with a bottle of champagne before a crowd of two thousand onlookers. One month later, the California State Board of Health lifted the beach quarantine for ap-proximately two miles between the pier in Venice and Ballona Creek.

The honeymoon did not last long. In November 1950, buoyed by his strong showing in the 1949 campaign, Aldrich led a recall movement against

Mayor Bowron, placing his own name on the ballot to replace the incumbent mayor. Although one might wonder where he found the time to run the Bureau of Engineering, Aldrich seemed to thrive on controversy. During the year in which he waged his recall campaign, his staff began the complex and arduous task of testing and operating the new plant section by section while political thunderstorms continued to roll over their heads. Aldrich captured 270,000 votes in the November election, but failed to unseat Bowron, who easily defeated the recall with 431,500 votes. Aldrich's 1937 civil-service protection came in handy, and he returned to his duties as city engineer.

By April 1951, the treatment plant at Hyperion was able to provide full primary and secondary treatment to the incoming sewage. In May, in a booming economy, voters approved an additional twenty-one million dollars for sewer improvements. Conditions on the beaches improved measurably, and in July the California State Board of Health removed all quarantines. But by summer, operational difficulties became the basis for a remarkable series of hearings that were reminiscent of some of the congressional hearings then creating headlines throughout the country as congressmen probed for communist subversion in the nation's government, schools, labor unions and the motion picture industry.

For sixteen days Mayor Bowron sought the cause of the "many bugs" in the plants. He announced his dissatisfaction with his own Board of Public Works. The board was reluctant, as Bowron had requested, to move responsibility for operations of Hyperion to Warren Schneider, then director of the Bureau of Sanitation. Both Franklin Thomas and Aldrich's longtime foe A.M. Rawn testified at the hearings and chided the city for placing its treatment plant so close to the beach. They also derided the 1-Mile Outfall, claiming that a longer outfall with a lesser degree of treatment would have sufficed.

On October 11, the Mayor stepped up his campaign and, in a hearing held in his office, accused the Hyperion plant of being a major contributor to smog. Bowron attacked the city engineer, claiming that Aldrich knew this and told no one and was fiscally irresponsible to boot. Aldrich was present during the entire presentation and had no comment as he sat smoking cigarettes throughout the hearing.

A week later, the Board of Public Works did what the mayor wanted: supervision of the Hyperion Treatment Plant was transferred from the Bureau of Engineering to the Bureau of Sanitation. *Southwest Builder and Contractor* magazine claimed that the move put "street sweepers" in charge of a "complicated and delicately balanced sewage treatment plant" and that "the city is apparently being penalized in this serious and dangerous way because the City Engineer, Lloyd Aldrich, had the temerity to run against Bowron."

After a drought of four years, the winter of 1951-1952 produced a near-record twenty-six inches of rain. At the same time, the city was experiencing another of its periodic booms. Almost one million people had moved into the city between 1940 and 1950. The results in the sewer system were inevitable. There was more water than the system could handle. Sewage spilled from pumping plants and from manholes throughout the city, but especially in the San Fernando Valley.

Frank Wada was a young chemist working at Hyperion. In 1994, he recalled that television viewing, a relatively new phenomenon at the time, affected the sewer system:

> When *I Love Lucy* was on and there would be a commercial break, there would be a sudden need to use the bathroom and you would see a big blip in the flow. The immediate remedy for the sewer maintenance crews was to take a truck and plant it on top of the manhole that was popping. This only resulted in sewage backup into people's bathrooms. So, the solution was to release the sewage to the Los Angeles River.

Two years earlier, the California State Legislature enacted a far-reaching measure, the Dickey Water Pollution Act. The legislation created nine Regional Water Pollution Control Boards that would be responsible for protecting surface, ground and coastal waters in their regions. Each of the Regional Boards would be semi-autonomous and could issue protective and enabling orders to public and private agencies. The California State Board of Health would no longer supervise or issue permits for discharge of sewage.

In 1952, the Los Angeles Regional Board began keeping records on the city's performance. These records show frequent instances of raw sewage spills during the 1950s. A serious health problem developed in the community of Reseda in 1952, as sewage flowed from several manholes into city streets. To protect the public health, the Bureau of Sanitation diverted sewage into the Los Angeles River. A portable chlorinator was used to disinfect the wastewater as it passed into the river, but it was insufficient to solve the problem.

In March the California State Board of Health commissioners wrote to the Regional Board to protest the growing unsightly sludge beds lining the river. In June a public health engineer surveyed the riverbank, which now "held heavy accumulations of black, decomposing organic material from Reseda to Balboa Boulevards, a distance of two miles." The board expressed concern about children who played in the river. The sewage problems in the San Fernando Valley would not be resolved for several years.

Mayor Bowron, who had served for fifteen years—longer than any previous mayor—was again running for office. The mayoral election in 1953 took place at the height of a conservative, postwar era that David Halberstam described as "a mean time," a time of widespread witch-hunts. Congressional, senatorial and state hearings investigated and branded as disloyal, ordinary citizens, popular entertainers, labor unions and political figures. Republican Dwight Eisenhower swept into office. The independent and liberal Republican Mayor Fletcher Bowron, who had initiated a series of reforms in city government that eliminated corruption spawned by his predecessor Frank Shaw, also sponsored public housing in Los Angeles, which outraged many of his former supporters. He faced widespread opposition from conservative forces in Los Angeles, who convinced Norris Poulson, described by historian Tom Sitton as a "bland but dependable congressman," to challenge Bowron.

Poulson was backed by the *Los Angeles Times*, the Hearst newspapers and a group of prominent lawyers including Frederick Dockweiler, founder of the Committee Against Socialist Housing. Dockweiler was the nephew of John Henry Dockweiler, who built the first outfall to the ocean in 1894. The beach adjacent to the Hyperion plant was named Dockweiler Beach in 1947 in honor of Frederick's father, the prominent Democratic Party stalwart Isadore Dockweiler. Isadore was the last Dockweiler to be associated with either the sewers of Los Angeles or the beaches onto which the sewage spilled.

Lloyd Aldrich made his third and final bid for the office when he entered the race in January 1954. He came in third in the primary behind Bowron, who was forced into a runoff with Poulson, who defeated the liberal mayor in the June election.

The world's finest and largest sewage treatment plant was enjoying a brief period of acclaim. Popular novelist Aldous Huxley celebrated the "miracle of Hyperion," when he wrote a twenty-two-page essay extolling the virtues of the new plant. His essay was one of many published in 1956 in the collection *Tomorrow and Tomorrow and Tomorrow*. Huxley recalled that in 1938, he and Thomas Mann and their wives were strolling along the beach south of the Hyperion Screening Plant when they suddenly discovered what he described as hundreds of white balloons. As he realized that they were looking at condoms, the hiking party also became aware of the nauseating smells surrounding the beach area and fled the polluted sands. After he visited the plant in 1953 he wrote, "What a joy the place would have brought to those passionately pro-saic lovers of humanity, Chadwick and Bentham!" alluding to the illustrious nineteenth-century sanitarian Edwin Chadwick and English philosopher and reformer Jeremy Bentham. Huxley wrote:

> As the fecal tonnage rises, so does the population of aerobic and anaero-bic bacteria. The chemical revolution begins in a series of huge shallow ponds, whose surface is perpetually foamy with the suds of Surf, Tide, Dreft and all the other monosyllables that have come to take the place of soap . . . The problem of keeping a great city clean without polluting a river or fouling the beaches and without robbing the soil of its fertility, has been triumphantly solved. The Hyperion Activated Sludge Plant is one of the marvels of modern technology, its effluent purity of 99.7 per-cent exceeds that of Ivory Soap.

Huxley's enthusiasm regarding the suds of the new detergents was not matched by either El Segundo residents or by plant personnel. Detergents that produced large amounts of suds were particular favorites of homemakers who believed, as many still do, that soapsuds connote cleaning. Unfortunately, the primary ingredient in these new sudsy detergents was the petroleum derivative ABS (alkyl benzene sulfonate). ABS is a complex molecule that is not readily decomposed by bacteria in the process of sewage treatment. The detergents, while cleaning better than soap, produced suds that did not biodegrade. Foam then formed on the surface of the water. The results were inevitable. Shortly after the plant was in full operation, billows of foam, often six to eight feet high, floated up and over the hill to El Segundo. Chemist Bill Garber, then in charge of the city's laboratory at Hyperion, later recalled, "The billowing foam that floated over the hill also carried grease and we found through our studies that the pathogenic bacteria tended to be brought to the surface, and they would also go over with the foam."

Even worse for the small community were the smells coming from the plant. Concentrated odors from the open headworks building (containing the bar screens and grit chambers) and smells emanating from the open aeration tanks and the gas collected from the digesters produced a stench that was often overpowering.

Despite the town's concerns, engineers outside El Segundo were enthusiastic about the plant. The American Society of Civil Engineers listed the Hyperion plant as one of the seven wonders of civil engineering in the Los Angeles area. Hundreds of foreign and U.S. engineers visited the plant during 1953. Industry publications praised the "unsurpassed achievement in modern industrial buildings." They warmed to the color schemes of the exterior buildings and gushed over the interior colors and landscaping, "designed to blend with the surrounding sand dunes." (Thirty years later, the plant was run-down and the landscaping had vanished. However, two water ponds were installed in the late 1970s and filled with the treated effluent. Dozens of turtles and many koi lived in these ponds, fed by an ever-attentive laboratory staff at the plant.)

The production of fertilizer—a key feature of the new plant—was not working out. Engineers were unable to filter out abrasives in the sewage, and a lot of the sludge was washed away into the channel that carried the treated wastewater to the 1-Mile Outfall. The water quality of the discharge was not meeting State Water Pollution Control Board standards. Also, the market for sludge fertilizer was pretty well cornered by the Los Angeles County Sanitation District, which had years earlier signed a contract with Kellogg to produce the popular Nitrohumus product. City engineers would later insist that the city's inability to sell the sludge as fertilizer led to major changes in treatment at Hyperion in 1956. Other reasons for change would soon become apparent.

On January 9, 1954 the California State Board of Health ordered the city to stop dumping sewage into the Los Angeles River by June 1. Mayor Poulson issued a call to San Fernando Valley residents to conserve water and to change their hours of doing the family wash to help alleviate overloading of the sewer system. A plan to stagger washdays was abandoned in March when too many people opposed the project.

As an interim solution to the overflowing sewers in the Valley, in August 1954, the city built a settling basin near Griffith Park to draw off excess sewage from the North Outfall while construction was started on a La Cienega and San Fernando Valley Relief Sewer. The aboveground Settling Basin was located next to Travel Town, at Crystal Springs and Riverside Drives. With two holding tanks, the basin was one hundred yards long and thirty feet wide. A year later, Metcalf & Eddy recommended building a sewage plant on Colorado Boulevard near Griffith Park, adjacent to the Los Angeles River. That facility, the Los Angeles-Glendale Plant, was eventually built in 1976.

In 1955, Frank Wada was a young chemist attempting to solve the problems of capturing sludge from the treatment process at Hyperion when he was sent to the new Valley Settling Basin in Griffith Park. His task: find the means of coagulating the solids, so they could be removed and pumped back into the sewer system during low flow, usually between 2 a.m. and 6 a.m. "We gauged the flow at the low spot in Burbank," Wada recalled, "and when it reached a critical point we'd start drawing wastewater into the basin" Ferric chloride was used at first but, Wada said, "being acidic it is reddish in color and turned the river red, so we turned to using alum, which, being white, was quite satisfactory." He and his colleagues monitored the river and there were no adverse reactions to the operation. The water was pumped to a chlorine contact tank for a couple of hours and then pumped into the river.

The Regional Water Pollution Control Board monitored the activity and seemed to find it acceptable. The Valley Settling Basin was designed for use for no more than a year or two while two new major interceptor sewers were being completed. However, because of other difficulties, the basin was activated several times in the following years, until it was phased out in 1959.

As it became clear that the new treatment plant could not adequately treat the more than 250 MGD then pouring into Hyperion, Aldrich suggested that the city spend an additional forty million dollars to improve and expand the facilities. However, Los Angeles County Engineer A.M. Rawn had other ideas. He submitted a report in February 1954 that outlined possible areas east of the Los Angeles River from which he proposed to divert to the county's system. He claimed the county could then handle one-third of the city's wastewater. Rawn and Aldrich soon engaged in public debate. At one point, in a joint interview with the editorial board of the *Times*, Rawn claimed that his plan for diverting 127 MGD from the east drainage area would save the city of Los Angeles fifteen million dollars in capital outlay and compared the county's operating expense of three dollars per million gallons to the city's $18.15 at Hyperion. Aldrich countered that the waters at the county's outfall site near White Point were contaminated and that kelp beds had been destroyed and commercial and sport fishing threatened. He went on to counter Rawn's cost analysis as a work of "fantasy." Rawn's plan was not adopted by the council.

In August 1954 Mayor Poulson appointed a twenty-one-member Citizen's Committee on Sewerage that included representatives from the California Department of Public Health, the city engineer's office and experts in the sanitation field including San Francisco consulting engineer Clyde C. Kennedy and Pasadena engineer Richard Pomeroy. The committee was asked to draw up a feasible and practical alternative to the current inefficient and costly operations at Hyperion.

With the committee's blessing, the city paid the Allan Hancock Foundation to conduct oceanographic surveys with bathymetric charts, sediment studies and topographic details of the ocean floor. The Board of Public Works obtained analytical reports from the U.S. Navy Electronics Laboratory and the Institute of Marine Resources at UCLA. LeRoy Crandall Associates made onshore geological surveys, and the Scripps Institute of Oceanography was hired to review and comment on the general oceanography program.

The direction and velocity of bottom flow, areas of erosion and sources of organic sediment in Santa Monica Bay were evaluated. Studies of wave action, marine life and bacteriology were used in conjunction with the examination of salinity distribution and vertical water temperature to determine where to put two new pipelines in the ocean. The report and accompanying technical data were sent to the council and Mayor Poulson on December 17, 1954.

The committee recommended a major change to the way in which Hyperion was treating the sewage. They said two new lines should be built. A 5-Mile Outfall would be used to carry the treated water to sea. A second line, a 7-Mile Outfall, would be built to carry all the sludge farther out and deposit it into a 350-foot cavern at the end of the Ballona Creek outlet.

Equally important, the consulting engineers said that the treatment time should be reduced from about six hours to little more than one hour. So much sewage was pouring into the plant that the six-hour detention time was too long—most of the sewage was not being adequately treated. Moreover, the long detention time was causing backups in the system. In addition, the committee included plans for modifications to several buildings and to the final settling system.

The committee claimed that when modified, the plant would be able to handle an anticipated 450 MGD in the decades to come. So, instead of giving all the sewage secondary treatment, only 100 MGD would receive the advanced treatment of aeration and final clarification. The rest would receive only primary treatment. The primary-treated water would be mixed with the smaller amount of secondary and would then go out the 5-Mile Outfall. The one-mile line built earlier would be retained as an emergency line in case pumps failed, or during heavy rainfall when too much mixed storm water and sewage poured into the plant.

In September 1954 the city purchased its first boat, the *Prowler*, a converted fishing trawler. The trawler would be used around the new outfalls to collect fish and water samples that could be taken back to Hyperion and analyzed by scientists in the laboratory there.

With the support of all concerned, a bond measure for sixty million dollars was placed on the April 1955 primary election ballot. It was an all-inclusive measure calling for the construction of two new ocean outfalls, improvements at Hyperion as suggested by Kennedy and Pomeroy, and construction of a new major interceptor line—the North Central Outfall Sewer—to relieve the burden on the old Central Outfall built in 1907.

In 1955, the city was experiencing a wave of optimism, prosperity and growth. Home prices were reasonable and jobs were plentiful. "The Ballad of Davy Crockett" topped the popular music charts and the automobile was king as commuters enjoyed the luxury of driving through the four-level interchange on freeways completed only two years earlier. Most paid cash for their new two-toned cars, while tourists visited Knott's Berry Farm and awaited the grand opening of Disneyland. However, of the two million residents, less than twenty percent bothered to vote. Fortunately, those who did enthusiastically supported the sewage measure: 340,302 votes in favor, 37,406 against.

In 1955, Mayor Poulson appointed Retired Admiral Cushing Phillips to the board. "I wanted to get a retired construction engineer to head the board and rein in Aldrich," he recalled in 1966. Aldrich retired that year, at seventy, after Poulson gave the city engineer a sweetheart consulting contract for three years. The generation of city engineers who followed Aldrich would never dominate city construction nor attempt to enter the political milieu as he had.

Equally significant, no one would ever be able to override either public opinion or the wishes of a mayor in office. Aldrich's activated-waste treatment plant meant that the city would always have to rely on Hyperion as the primary plant to handle the city's wastewater. In the 1930s and 1940s, the city could have purchased land near the Baldwin Hills, as well as a second site east of downtown, close to the Los Angeles River. Because Aldrich refused to consider that possibility, though the city eventually built two upstream plants—one near Glendale and the other in the San Fernando Valley—those plants relieved less than one-fourth of the city's sewage in the coming decades. As it does now, Los Angeles then treated most of its sewage at one site: Hyperion.

Shortly after Aldrich retired, Joe Nagano, manager of the Hyperion Laboratory, recalled that the times after the plant was finished in 1950 were truly halcyon days—not just for him, but for many of the engineers and operators:

> It was a beautiful place. The marine phenomena were quite incredible—at times, just millions and millions of fish, you know washing onto the sand. One time I saw the ocean black with the bonito. They were side by side from the shore—clear to the horizon. There was a tremendous sardine and anchovy fishery there and the fishing trawlers would net the fish every single day. They would never run out of fish.

This picturesque scene provided a brief interlude as conditions would change in the coming years. The plant reduced its treatment of sewage, and the view from the Hyperion Laboratory was lost when a new administration building blocked out those dreamy vistas.

CHAPTER TEN

An Interim Solution to Pollution

Lyle Pardee began work with the City of Los Angeles as a draftsman in 1923, when he was twenty-one. He was chief deputy city engineer when the Board of Public Works appointed him to the post of city engineer in December 1956. While he was in charge of all public works projects during the next sixteen years, the city faced major sewage challenges. Engineers who worked under Pardee recalled that he was slim, probably no more than five feet, eight inches tall. A city photograph shows a man in large, square eyeglasses, sporting a 1930s-style moustache, a man who—in both appearance and attitude—was completely different from the florid, expansive and politically driven Lloyd Aldrich. Council and board records evidence a competent, even creative engineer, but not an imposing man. Although he enjoyed the perks of office, including a chauffeur, Pardee was not a man to seek the public spotlight.

Perhaps he was uncomfortable with the emotions that political conflict could arouse. Pardee was an accomplished transportation engineer who designed and supervised construction of the Arroyo Seco (Pasadena) freeway and participated in the design and construction of other city freeways built in the 1950s and 1960s. He supervised all construction of the Civic Center, including Parker Center, headquarters for the Los Angeles Police Department.

Pardee received high honors from the engineering profession, including being named one of the top ten engineers in the country by the American Public Works Association in 1963. But he had little to do with the sewer system improvements while he was city engineer. Though he was ostensibly in charge of the changes recommended by the 1954 Board of Engineers, these improvements were carried out by a team of construction contractors who called themselves the Hyperion Engineers. City engineers and a consulting board of engineers appointed by Mayor Poulson reviewed and approved all the plans and work, but the actual construction of the North Central Outfall, the changes and improvements at the plant and building of the 7-Mile and the 5-Mile ocean outfalls, were completed under the supervision of the contractors. David L. Narver, from the internationally renowned construction firm Holmes & Narver, was project manager for the North Central Outfall. E.H. Graham, Jr., of Koebig Sons was responsible for the modifications at the plant, while the design and architectural firm Daniel Mann Johnson and Mendenhall (DMJM) built the two new ocean outfalls under the supervision of Irvan Mendenhall and David Narver. Narver, who was the actual supervising engineer for all these projects, later recalled that he worked in close coordination with many city engineers, but did not work with City Engineer Lyle Pardee.

Of the two ocean outfalls, the longer 7-Mile sludge outfall was completed first, in October 1957. That's because the whole system of filtering and drying the sludge to create fertilizer was not working well at all. After digestion, the sludge had to be washed to remove alkaline components and ammonia before it could be filtered. Unfortunately, the sludge was so fine that particles of it were washed out in the process and wound up in the same effluent that had previously received full secondary treatment, making the discharge less than adequate. Most of this sludge came back and settled on the ocean floor between the end of the outfall and the shore. So, the engineers wanted an alternative disposal plan to be implemented as quickly as possible.

The new 7-Mile sludge line would deposit the sludge several miles from shore in a three-hundred-foot-deep cavern, directly west of Ballona Creek. The line was a twenty-inch steel pipe, coated on the outside with gunite, and cement-lined on the inside. A pumping plant was built so that the sludge, often less than four percent solids, could be mixed with additional water for ease in moving it from Hyperion through the 7-Mile line. Although the city began using the new sludge line in 1959, fertilizer production continued until August 1960. The city now discharged 135 tons of sludge a day into the ocean; this essentially killed most marine life in and around the cavern.

The approximately seven million dollars of equipment used in washing, filtering and drying the sludge was decommissioned, but the buildings that housed various aspects of production remained on-site until well into the 1990s. These buildings were especially popular with Hollywood location scouts in the late 1960s and early 1970s. *Planet of the Apes* (1968), *Soylent Green* (1973) and other movies used these buildings and other parts of the plant to establish the menacing quality common in science-fiction films of the era. A noxious air-polluting tower used in part of the sludge-drying process was eventually torn down on April 26, 1977.

In 1957, David Narver and his team began working on construction of the North Central Outfall. This outfall was built to relieve the pressure from the Central Outfall. Eight miles long and twelve feet in diameter, it was constructed in two sections and was completed in September 1958. It is a circular reinforced-concrete pipe. Some of the concrete was precast. Some was poured in place and then reinforced. The bottom of the sewer had a steel-trowel finish, to improve resistance to abrasion from the sewage. Although some portions were laid in open-cut trenches, the majority was laid in tunnel at a depth of 190 feet. The upper 260-degree section of the outfall pipe was lined with vinyl plastic, to protect the concrete from hydrogen sulfide gases emitted from the sewage. The new outfall, which is still in use, extends from the intersection of La Cienega Boulevard and Rodeo Road to the north end of the Hyperion Treatment Plant.

The North Central Outfall does not discharge directly to the ocean as did the original Dockweiler outfall and its replacement, the Central Outfall. Engineers call sewage pipes outfalls when they carry sewage to a disposal point at the ocean or into the ocean. The North Outfall of 1925 ended at the Hyperion Screening Plant, where the new 1-Mile Outfall was built to convey the sewage from the plant to the ocean. The North Central Outfall ends at the Hyperion Treatment Plant, but the new 5-Mile and 7-Mile ocean outfalls begin at the plant. After the North Central sewer was finished in September 1958, the hydraulic capacity of the three main outfalls leading to Hyperion exceeded the capacity for full treatment at the plant. So, the city ended full secondary treatment and the plant converted to standard-rate activated sludge treatment. Secondary flow was lowered to 100 MGD, and the rest of the sewage received only primary treatment. The mixture of both primary and secondary effluent, after chlorination, was now discharged out the 1-Mile line, in violation of the city's state discharge permit—a condition that lasted until the new 5-Mile Outfall was completed in 1961.

The 5-Mile Outfall was designed to disperse the combined primary and secondary effluent away from the beaches. Twelve feet in diameter, the outfall pipe terminates in water approximately one hundred ninety feet deep. At the end of the pipe, two diffuser pipes that are each four thousand feet long were added in the shape of a "Y" with each diffuser going off at a sixty-degree angle. Six-inch diffuser ports (holes) were spaced every six feet along the ocean side of the pipe, for a total of 667 per diffuser leg, or a total of 1334 diffuser ports. Diffusers work just like an irrigation system on a lawn. The effluent water is pressurized in the pipe, so it flows out equally at each port. To keep the flow distributed, the diffuser is tapered, starting at six feet in diameter, shrinking to four feet at one-third the way, then to two feet at two-thirds the distance from the main trunk. The diffusers disperse the treated sewage over a wide area, reducing the impact of pollution.

A new effluent pumping plant housed five pumps to handle primary-treated effluent. Like many aspects of Hyperion, the pumping plant was one of the largest ever designed. It was designed to handle a dry-weather flow of 600 MGD and should have handled a storm flow up to 720 MGD. Although the entire flow from the plant could flow by gravity during many hours of the day, some pumping would be required when the flow exceeded 300 MGD. The pump station was entirely automatic, so it was vulnerable to power outages, especially in the 1970s and 1980s when excess primary effluent would be diverted out the 1-Mile—sometimes chlorinated; sometimes not, depending on the alertness of the shift operators at the plant.

In May 1956 the California State Board of Health had introduced new standards and rules designed to protect the beneficial use of the water in Santa Monica Bay. The city now was required to establish twenty-four sampling stations in a monitoring program to include grab samples of surf water for a distance of fourteen miles north and nine miles south of the Hyperion plant. The program was put into operation in July that year.

Following the successful completion of the three new outfalls, four new primary tanks and the new effluent pumping plant were added at Hyperion in 1960 to increase treatment capacity at the plant. By September 1961, all the plant effluent—combined primary and secondary—was diverted to the 5-Mile Outfall, and the effluent pumping plant was placed on automatic operation. The 1-Mile remained available to be used for overflows or when the automatic pumps failed.

Disquiet about pollution of America's air and water was growing throughout the country. As early as 1960, many American cities were reporting excessive foam not only at sewage treatment plants but also in water supplies. Alkyl benzene sulfonate (ABS), one of the many chemical engineering marvels of the mid-1950s, was not only causing disruption in municipal sewage treatment plants, but was turning up in drinking water as it permeated groundwater tables, rivers and streams.

In 1963, California State Assemblyman Don Allen introduced a bill that would have banned detergents that contained ABS. The city had been forced to install a special grease-skimming system at a cost of more than $350,000 at Hyperion to combat grease that was attaching itself to the persistent ABS. Allen's bill failed, but eventually, in 1967, legislators and sewage engineers convinced detergent manufacturers to switch from ABS to a more degradable synthetic, linear alkylate sulfonate (LAS), creating a "soft" detergent that resulted in lower suds output.

Although ABS was never considered a health hazard, other miracle products of the chemical age were producing increasing evidence of major pollution of the air and the sea. Water pollution became tied to these chemical wonders, especially pesticides and herbicides. The first major awakening to the visible menace was provided by the biologist Rachel Carson. *Silent Spring*, Carson's indictment of pesticides and herbicides that were destroying the balance in the oceans of the world and causing widespread destruction of insect and bird life, awakened Americans to the deadly dangers of DDT. Published in

1962, her book caused a national debate on the use of chemical pesticides. Her detailed, extensive scientific analysis of the toxic legacy, carcinogenic properties and extended life of DDT became the foundation for a federal ban of the pesticide in 1972.

The largest manufacturer of DDT was Montrose Chemical Corporation. From 1947 to 1971, the company manufactured DDT at its Torrance plant on South Normandie Avenue. Beginning in 1953, it was permitted to discharge DDT-contaminated wastewater into the sewer system. For almost two decades before monitoring began in 1970, an unknown but large amount of DDT—estimates range up to 1800 tons— traveled thorough municipal sewers before emptying into Santa Monica Bay near the Palos Verdes Peninsula. DDT remains a potent poison on the ocean floor of Santa Monica Bay well into the twenty-first century, and Angelenos are still frequently reminded not to eat bottom- feeding fish, particularly white croaker. As late as 2007, DDT levels in these fish remained five times higher than found anywhere else in the world. Some DDT did travel through the Los Angeles sewer system, but the source of most of the sewer-borne DDT was the Los Angeles County Sanitation District's White Point Outfall off the Palos Verdes Peninsula, with some contamination off Cabrillo Beach from the city's Terminal Island Treatment Plant.

The new improvements at Hyperion were extremely important because the San Fernando Valley population was exploding in the rapid-growth era of the late 1950s. The Valley's population grew to more than 739,000 in 1960 from only 112,000 in 1940. No other part of the city had grown so quickly. Developers and contractors were scrambling to build housing for the expanding suburb. It was the third-largest population boom in the city's history. This time, the engineers were determined to provide the necessary public works projects to serve these communities. The La Cienega-San Fernando Valley Relief Sewer, completed in 1955, provided only partial relief for the Valley.

"You had the whole sewer system of the Valley being converted from cesspools in some parts of the Valley to new subdivisions that had to reach out with interceptors to connect to the system," retired City Engineer Donald Tillman recalled. "So that's how I became involved in the sewers. And wastewater was not my thing. I didn't think I'd end up there, but before it was over I was into it pretty deep."

Donald Tillman first came to work for the city in 1947 after serving in the Civil Engineer Corps of the United States Navy during World War II. He spent the first four years with the Bureau of Engineering working on freeway design but decided to join the Bureau of Sanitation when an opportunity for advancement opened up. A few years later he moved back to the Bureau of Engineering, working in the San Fernando Valley where huge development was taking place. It was there that Tillman developed a strong relationship with contractors and developers. These men brought him to the attention of Mayor Poulson and his longtime chief ally, City Administrative Officer Samuel Leask. The CAO— an appointed position—is the financial advisor to the mayor. He has the key responsibility for preparing and recommending all city department budgets, almost always adopted by the mayor before the complete city budget is presented to the City Council for final approval.

In 1960, while the sewage problems at Hyperion appeared to finally have been resolved with the interim improvements, Poulson had been attempting to reduce the power of the Board of Public Works. CAO Leask was working on a report that would recommend replacing the board with "a strong professional administrator," to manage the complex responsibilities of the several separate bureaus which then employed more than six thousand workers with a collective budget of fifty million dollars. Poulson's effort to reduce the power of the board was only one of several such attempts, beginning in 1949 and continuing to the 1990s. Despite editorial support for such reforms from most of the city's newspapers and such venerable organizations as the League of Women Voters, the public consistently refused to dilute the board's power.

When Mayor Poulson announced his plans to revamp the responsibilities of the board in January 1960, board president Admiral Cushing Phillips submitted a letter of resignation. Poulson prevailed on Cushing to postpone his retirement for some months while he and Leask sought a replacement. Since its formation in 1906, a civil engineer had always been a member of the Board of Public Works. Leask and Poulson selected thirty-five-year-old Senior Civil Engineer Donald C. Tillman to replace Phillips.

When asked about his appointment, Tillman later said that he had never been told why. Tillman acknowledged that "Samuel Leask had something to do with my appointment," but insisted that he was chosen because Leask and Poulson could not "find another Navy admiral, who was retired and willing to work for a relatively small salary [commissioners were paid $13,700 a year and Tillman's salary at the time was $12,576]." Although not an admiral, Tillman was a lieutenant commander in the Naval Reserve. A native Angeleno, he was broad-faced and handsome, a husky athletic man—six feet, four inches, 245 pounds, who was known to have a strong ego. At thirty-five, he would be the youngest member of the board and the first and only city employee promoted to commissioner. Tillman became the newest board member in July 1960.

The Board of Public Works, with its five commissioners, continues to be the only full-time salaried commission in the city, as it has been since its first meeting in 1906, when commissioners were paid an annual salary of three hundred dollars and were required to post bond to ensure their honesty in awarding contracts. The commissioners have long since been allowed to serve without posting bond. Their autonomy and power are underscored by the visual splendor of their meeting room at City Hall. The members meet three times a week, in public, in a large hearing room at City Hall. Although it is almost half the size of the City Council meeting room, the two-story-high board meeting room, with walnut pews, marble pillars and an Art Deco ceiling, is an extraordinary meeting place. Elaborate chandeliers cast a warm glow over an often-intimidated audience of appellants, bidders and city employees. The board members sit on a raised platform in large leather armchairs around a massive walnut table, on which sit stacks of documents to be reviewed.

The board's secretary sits at the table as well, wielding the power of continuity and integrity. By 1960 the job had been held by only two men in the entire fifty-eight-year history of the board. Members of the board came and went, serving at the mayor's pleasure, but Horace B. Ferris, appointed in 1906, who served until 1939, and Milton Ofner, who succeeded him, kept the calendar for the board, organized the agenda, presided over and collected the public bids. The secretary conducted daylong briefing sessions for each new member, outlining duties and responsibilities and maintained detailed handwritten records of all board meetings, subsequently bound in oversized yearly journals

The board president, however, directs the meetings. The members, who rarely disagree in public, wield power and influence in local politics. The five-member board holds assessment hearings, determines the purchase and sale of public property, issues and awards contracts for all public works projects and in the 1950s and 1960s was comprised exclusively of white, Anglo-Saxon males who would vote favorably on development issues and would provide political support for whomever was elected mayor. These jobs were sought-after political plums for the mayor to hand out.

Norris Poulson had planned to retire in the winter of 1961, but was persuaded by Republican Party leaders and a coalition of business interests to run for a third term. Although Poulson faced an extensive list of contenders in the primary election, one candidate quickly stood out from the pack: Sam Yorty was by any measure a determined candidate, one who had already earned the well-deserved sobriquet of "constant candidate." His career encompassed a few hits and even more spectacular losses. His record included three earlier entries into the race for city mayor: the 1939 recall election of Frank Shaw, when he ran a distant sixth, and a similarly dismal record against reform mayor Fletcher Bowron in 1941 and 1945. He was, however, elected to the State Assembly in 1936 and re-elected in 1943, after which he received his law degree. He then served two terms in the U.S. Congress from 1951 to 1955. Yorty lost a senate campaign against Hiram Johnson in 1940 and was the unsuccessful Democratic candidate for U.S. Senate in 1954 against Republican Thomas Kuchel. Not to ruin his record as frequent loser, he would later run unsuccessfully for governor in 1965 and 1970, as well as for the presidency in 1972. But 1960 was Sam Yorty's year.

He announced his candidacy in February and immediately formed what he described as a "Truth Squad," to counter incumbent Poulson's earlier creation of a "Facts Brigade." He claimed that Poulson was backed by a "powerful and ruthless" political machine and alleged that Poulson had a "slush fund" of two hundred fifty thousand dollars to his meager five or six thousand dollars. Yorty presented himself as the honest underdog in a race that most observers assumed would be won by the bland but honest Poulson. The mayor's Facts Brigade, however, was no match for the indefatigable Yorty, who was a frequent guest on television and radio. The Aldrich-Bowron battles in previous years were gentlemen's disputes compared to the Yorty-Poulson war. When Poulson claimed that Yorty was supported by "underworld figures," Yorty immediately filed a slander suit against the mayor. He kept slugging, forcing Poulson into a surprising runoff after placing second in the primary.

The annoyed and beleaguered sixty-five-year-old Poulson, who suffered an unrelenting case of laryngitis and appeared sick and wan on television, was attacked constantly on radio and television by the freewheeling, pugnacious and clearly healthy, always smiling, younger, fifty-two-year-old Yorty—who had a head of thick, dark, wavy hair and had wisely discarded an earlier heavy moustache. During the campaign, Yorty seized on a key issue that had been troubling the city for some time. Housewives had been forced to separate their trash for collection in a complicated system. City crews picked up garbage twice a week, which was then delivered to hog farmers. Grass clippings, paper and other combustibles were picked up once a week and then delivered to landfills. Tin cans and other metals were collected once every three weeks and then delivered by city crews to a private salvage company that had a contract to buy the material from the city.

In his campaign Yorty grabbed the trash issue and threatened to "foment a housewives' revolt" after he was elected if the full council did not approve combined collection. Although Yorty would go on to promote sewage improvements and was probably the best-informed of all city mayors about sewage matters, he would forever be enshrined in local memory as the man who established a combined collection of trash and saved homemakers from the onerous task of separating their garbage. (Ironically, the city began a separate collection for recyclables in the 1980s after diminishing landfill space spurred the city into beginning a recycling program that would eventually divert almost seventy percent of the city's trash to recycling facilities.)

Yorty captured the majority of votes in the "nonpartisan" election by sixteen thousand votes, a slim margin. In a post- election analysis, the *Times* wrote that his appeal to disaffected voters in San Pedro, the southwest areas of the city and especially the San Fernando Valley—as well as to those who objected to the Poulson administration's arrangement for the Dodgers ballpark in Chavez Ravine and the new Bunker Hill development, and who sided with Yorty's objections to an eighty-five-thousand-dollar study on refuse collection—were the chief reasons for Yorty's victory.

Shortly after his election, Yorty replaced all but one of the members of the Board of Public Works. Tillman had expected Yorty to fire him, but he was retained. The four new board members were Louis Dodge Gill, president of his own San Fernando Valley engineering firm, straight-laced, do-right C. Irwin Piper, a retired FBI agent who would later become the City Administrative Officer, former FBI agent Howard Chappell and San Fernando businessman Ernest Webber. They all got along well with Tillman. Piper, elected board president, was an effective defender of its authority.

Tillman's leadership qualities were apparent in 1962 when the board members unanimously elected him president when Piper resigned to become the new CAO after Leask retired. During Tillman's two and one-half years on the board, City Engineer Lyle Pardee and his staff were in essence working for Tillman. It must have been a bitter pill for Pardee to swallow to have Tillman, who had only recently been a civil engineer in the San Fernando Valley office, now supervising him. And Tillman, who in his new job was earning fifteen hundred dollars a year less than the city engineer, was not above pettiness, at least in one instance. In 1962, Tillman accused the city engineer of "abusing his driver privileges," and ordered the removal of Pardee's chauffeur, requiring Pardee to get the board's permission to have a driver when he needed one. The two men were in disagreement also about plans to satisfy the sewage needs of the San Fernando Valley. Pardee wanted to build a new tunnel through the Santa Monica Mountains. Tillman favored building an upstream water reclamation plant instead. While on the board, Tillman discussed his idea for a plant in the San Fernando Valley with Louis Gill and Bureau of Sanitation Director Norm Hume, who agreed with the alternative but did not publicly challenge City Engineer Pardee.

Although proud of the fact that he had been the youngest board member ever appointed, Tillman said that he quickly tired of the politics required of him, especially as president. He was, he said, naïve in thinking that everyone in city government wanted to work together to solve problems. His biggest annoyance stemmed from press releases about public works projects. Yorty wanted to announce the projects and get credit. But the council members of districts in which work was being done also wanted credit. To solve the problem, Tillman ordered the board's public relations officer to draft two releases with the same information, one under the mayor's name and one under the council member's name. The releases were sent simultaneously to both. "You had to touch all the bases," Tillman said, "All of a sudden it had proliferated into a nightmare of influence of people who you had to convince that your ideas were good." He wanted to get back to design and construction, so in 1963 he resigned from the board and was appointed chief deputy city engineer—and now reported to Pardee. Since each became involved in different areas of responsibility, the two men did not air their differences. Pardee spent most of his time obtaining more than two hundred million dollars in recently established gas-tax funds for use in street improvements and construction. Although rarely recognized or rewarded by local politicians for his many achievements, Lyle Pardee was an expert in highway and street design, and received numerous national awards for his work. Indeed, no mention was made in the local newspapers when he retired in February 1972 after forty-nine years of city service.

In 1961, as sewage improvements became increasingly evident to residents, voters overwhelmingly approved a sewer-bond measure for forty-six million dollars by a three-to-one margin. The measure listed a wide variety of improvements including additional interceptors, relief and bypass sewers, as well as reconstruction of the Central Outfall Sewer. The following year, however, despite opening of the new North Central Outfall, city engineers knew that during heavy rainfall the North Outfall might be subject to excessive pressure—enough to cause the pipe to burst.

Under Pardee's supervision, the engineers designed an emergency bypass structure at Jackson Avenue at the much-abused Ballona Creek. The emergency structure was relatively simple. Costing $161,324, the sewer bypass consisted of a hundred-foot-long, six-foot-wide, reinforced-concrete pipe connected to the North Outfall by a concrete box structure containing two manually controlled gates. Construction began in the summer of 1963 and was completed in January 1964, and the structure was used during the next major rainfall. Chlorine tablets were dropped into the box to disinfect the sewage—however inadequate that would have been for overflows of many millions of gallons. The bypass would be used for the next twenty-five years, causing a major public relations disaster in 1985 when conservation and environmental organizations led a highly visible campaign to protest its use and subsequent contamination of the creek.

This temporary makeshift solution to overloading sewage in the North Outfall did not solve the issue of the lack of sewage relief for the San Fernando Valley. City Engineer Pardee began design of a huge hydraulic pipeline that would divert sewage from the North Outfall from the Valley to West Los Angeles. The pipeline would go through a tunnel under the Santa Monica Mountains and would reconnect to the North Outfall nearer to the Hyperion plant. Annual reports in the early 1960s describe the new additional Valley relief sewer (AVORS) as "one of the most ambitious sewerage design and construction projects ever undertaken by the Bureau of Engineering." Pardee's tunnel was planned to begin construction in 1972.

Tillman, who from his days on the board was convinced that the tunnel project was unnecessary, continued to work on his alternative plan. "As chief deputy I was able to start work orders for an alternative plan for an upstream plant in lieu of another big elliptical tunnel through the mountains," Tillman said. He received support from Board vice-president Louis Dodge Gill. Although design on the tunnel was more than fifty percent completed, Tillman started cost studies to show that an upstream plant would cost less than another major interceptor all the way through Hollywood. If Pardee's plan for the tunnel had been approved, the city would have eventually been forced to build a third tunnel to accommodate the continuing growth in the San Fernando Valley.

In 1967, Gill gave orders to abandon Pardee's tunnel and gave the go-ahead to Tillman's water reclamation plant. It would be located in the Sepulveda basin, a flood control area managed and owned by the U.S. Army Corps of Engineers. Pardee was forced to defend the new plan to City Council members and, despite Tillman's initial estimate of fifteen million dollars for the project, Pardee's more realistic estimate was fifty million. Nevertheless, water reclamation had become increasingly popular in the past several years as the city and county began intensifying programs to recover sewage water. Tillman was brilliant in his vision of perhaps the most unusual and elaborate plant eccentricity for a sewage treatment plant. He decided to add a Japanese garden with a lake and a meandering brook as a buffer between the plant and the neighboring suburban community.

Tillman emphasized planning for future shortages and uses for reclaimed water. When interviewed in 1993 about his days with the city, he recalled his message: "You don't call it a treatment plant, little brother, you call it water reclamation. You glamorize it. Sepulveda Water Reclamation Plant. Where? The Valley. It's right there, not against the houses. The land's there. You can go piggyback on a big Valley outfall sewer. Save the water, you can reuse it."

He first had to convince the City Department of Recreation and Parks, which had the lease for the land and controlled the golf courses in the Sepulveda Basin. He arranged to fly the recreation and parks commissioners and department head William Frederickson to Golden Gate Park in San Francisco, where treated sewage water was being run through the park. "Look, we can do this," he said. "We can even do a Japanese garden. We can make it aesthetic." He knew what he was talking about. Tillman was then enrolled in a Japanese gardening class at UCLA and he was able to talk to the recreation people with the fervency of a true believer. They approved the project.

The next hurdle was tougher. The Army Corps of Engineers refused to approve the area for the use as a treatment plant, citing the possibility of a hundred-year flood inundating the area. Tillman went to board member Gill, who in turn went to Louis Nowell, chairman of the City Council Public Works Committee. Nowell took Tillman to Washington to speak with key legislators there. A dead end. But Gill, a respected structural engineer and now public works board president, went to Mayor Sam Yorty. "That was the breakthrough," said Tillman.

In July 1969, Yorty carried the plan directly to President Richard M. Nixon. "You'd never believe it." Tillman recalled, "Mayor Yorty took President Nixon a booklet, just a black-and-white architectural rendering. He showed it to President Nixon, who called in the head general of the Corps of Engineers and said 'What's wrong with this? This is a good idea. Reclaim water.' And the general said, 'Well, Mr. President, I guess it could be done.' That did it. We came home and were ready to celebrate." The city also allayed the flooding concerns of the Corps of Engineers when it leveled off several acres below the site of the planned treatment plant to increase the flood capacity of the Sepulveda Basin.

Construction on the plant, however, was delayed for almost a decade as the city began a fifteen-plus-year battle to avoid full secondary treatment of its sewage. As the struggle ensued, California and U.S. Environmental Protection Agency officials refused to release government money that could provide 97.5 percent of sewage plant construction.

In April 1967, after twenty-two years of voter approval for sewer improvements amounting to more than one billion dollars, residents decided that enough had been spent on the sewers. So, despite the pressing need for additional sewer improvements, voters turned a blind eye to the increasing sewage menace and rejected a council-supported charter amendment that would have imposed a sewer service charge on all real property.

CHAPTER ELEVEN

The Clean Water Battle

M eanwhile, the American public was becoming increasingly concerned about water pollution. Pollution in the Passaic River killed thousands of fish. The Hudson River showed signs of dying. The National Wildlife Federation, the Audubon Society and the Izaak Walton League of America campaigned heavily for protective federal legislation. Bays and estuaries in Seattle, San Francisco, Delaware and Philadelphia were being choked by a wide variety of industrial and municipal wastes.

Los Angeles would feel the effects of the environmental age that opened on the morning of January 28, 1969 about five miles off the coast of Santa Barbara. A "blowout" of Well A-21 at Union Oil's Platform A offshore drilling rig in the Santa Barbara Channel spilled two hundred thousand gallons of crude oil into the ocean, creating an eight-hundred- square-mile slick that

◁ Bacteriologist reviews tests at new Hyperion Treatment Plant. circa 1965

befouled the ocean and washed onto beaches. Images of the oil spill and the sight of thousands of dead birds and hundreds of oil-soaked grebes, cormorants and pelicans galvanized the nation. President Nixon commented, "The Santa Barbara incident has frankly touched the conscience of the American people."

On February 17, only a few weeks later, stories appeared in the news media about the tragic death of Lake Erie. Raw sewage, human excrement, oil and chemicals and rotting algae formed in long piles, clogging the shorelines and turning the lake into pea soup. All of the Great Lakes were suffering. "At the height of this degradation," wrote Theo Holborn in *Our Stolen Future*, "huge, stinking mats of rotting algae covered the beaches of the lakes, while bays and rivers were awash in oil and industrial waste, and once-abundant bird and wildlife populations were collapsing." On June 22, the heavily polluted and odoriferous Cuyahoga River caught on fire near Cleveland, awakening even the dimmest of us. But it was the Santa Barbara oil spill, with its constant visual reminder of the cost of pollution, that was the single most galvanizing event for the country.

Even before the oil spill off the Santa Barbara coast, people at the California State Water Resources Control Board were developing new stringent requirements for all sewage dischargers in the state. Every day, more than a billion gallons of treated sewage effluent was dumped into the Pacific Ocean from the Mexican border to the Santa Barbara County line. John Parker, chief engineer of Los Angeles County Sanitation Districts, recognized that the new requirements would soon impact Southern California sewage treatment plants and began to organize a research program. The program was quickly endorsed and funded in March 1969 with a budget of $1,133,000. Known as the Southern California Coastal Water Research Project (SCCWRP), it became a joint powers undertaking of Los Angeles, San Diego, Orange and Ventura counties and the City of Los Angeles.

SCCWRP undertook a three-year study of the condition of sixty miles of the ocean waters from the Mexican border to the Ventura County line. Study results would provide the foundation for a contentious, decade-plus battle by Los Angeles officials over state and federal requirements mandating improved treatment of sewage discharges to the ocean.

In July 1969, the California legislature approved the Porter-Cologne Act, which established stringent standards for discharges into the ocean from industry and municipal wastewater treatment plants. A panel from the State Water Resources Control Board had issued a report in January outlining a series of new requirements for agencies discharging sewage into the ocean. Despite intense attacks by industry and business, the report was approved and only slightly modified as its recommendations were passed into law.

Shortly after California passed the Porter-Cologne Act, United States Senator Edmund Muskie began a two-year series of congressional hearings into nationwide water pollution. In 1970 President Nixon created the National Environmental Policy Act. The act established new limits on water pollution and approved four billion dollars for the improvement of water treatment facilities. Nixon also created the Environmental Protection Agency (EPA), which would become a major enforcement tool in the battle to clean up the nation's waterways. The money would help Los Angeles build new facilities, but the EPA and state regulators would withhold hundreds of millions of dollars in funding until the city changed its ways and finally stopped sending millions of gallons of sludge into the ocean and discharging minimally treated sewage into Santa Monica Bay.

The clean water battle began when Congress approved the Clean Water Act of 1972. Senator Edmund Muskie was the author of—and the driving force behind—the Clean Water Act. More than any other figure in American history, Muskie was responsible for cleaning up the nation's rivers and oceans and "almost single-handedly created a national consensus for protection of the environment," according to former PBS correspondent Charlayne Hunter-Gault. He did this by creating and ensuring passage of the Clean Air Act in 1970 and overhauling the Federal Water Pollution Act to create what has come to be known as the Clean Water Act of 1972. The Clean Water Act marked a revolutionary change in the nation's water-pollution regulations.

The goal of the act was to "restore and maintain the chemical, physical and biological integrity of the nation's waters." It called for zero discharge of pollutants into navigable waters by 1985 and fishable and swimmable waters by 1983. It demanded that all sewage treatment plants convert to full secondary treatment to remove at least eighty-five percent of suspendable solids and to monitor the biological oxygen demand (BOD) of wastewater. High BOD in water is a significant indicator of pollution.

In addition to requiring full secondary treatment, the law mandated that all wastewater agencies had to stop discharge of sewage sludge into the ocean and that every sewer-system user pay his proportionate share of opera-tion and maintenance costs. Within three months of the act's passage, each state would have to establish standards for water quality and begin implementa-tion of a plan to meet those standards.

Many of the requirements written into the new federal law were based on California's Porter-Cologne Act of 1969. The California act provided for several new features in water pollution control: tough civil fines of up to six thousand dollars per day for violations of pollution cease-and-desist orders; mandatory cleanup of pollution by violators, with full liability for cleanup costs; two additional members of the public with water quality experience added to each of the nine Regional Water Quality Control Boards; increased staff for both the state and regional boards; tighter controls against conflicts-of-interest among members of the regional boards; and new, tougher standards regarding wastewater discharges from industry and publicly owned treatment plants.

California's new regulations limited discharge of heavy metals, required the removal of all sludge from sewage and drastically reduced allowable limits for "floating particulate matter, grease and oil, suspended solids and settleable solids." The California State Water Resources Control Board held that cities would be responsible and liable for everything discharged into publicly owned treatment plants, and would have to establish industrial waste ordinances to ensure that industries would eliminate most of their pollution (especially heavy metals such as mercury, nickel and cadmium) before their waste reached the plants. All facilities would be required to comply with the new regulations by 1977. These provisions became the backbone for a National Pollutant Dis-charge Elimination System (NPDES) enacted in the Clean Water Act. Los Angeles would have to meet California discharge standards to obtain a permit to operate its wastewater plants. The city would have to apply for a new permit every five years.

According to Leon Billings, Muskie's chief of staff on the Senate Committee on Water Pollution, the carrot to the stick of the new regulations was money—lots of it—if states complied with the new statutes. The federal government would pay up to seventy-five percent of construction costs for technologically advanced treatment, and the state would kick in an additional twelve and one-half percent. This appealed to the City of Los Angeles, which would nonetheless try to circumvent some of the new law's provisions as it went after federal money to build the new water reclamation plant in the San Fernando Valley.

In 1971, voters approved sixty million dollars for construction, repair and maintenance of reclamation plants, treatment plants, sewers and the land on which to build the new facilities. The measure included twenty-one million dollars for the new reclamation plant in the San Fernando Valley. However, the city decided to apply for federal and state funds to underwrite most of the cost of the new reclamation plant and use most of the bond money for other projects. But the state and Paul DeFalco, the new regional administrator for the EPA in California, demanded that the city stop its discharge of sludge into the ocean within the next five years if it wanted to receive additional federal funds.

In 1972, Lyle Pardee retired and Donald C. Tillman became city engineer. The city was now discharging three hundred thousand tons of sludge per year into the ocean through its 7-Mile Outfall. Although city engineers did not really believe that the sludge in the ocean was an environmental hazard, they nevertheless developed a plan called "Desert Bloom" to send the sludge to the Mojave Desert. The idea was to pump sludge slurry to Edwards Air Force Base, creating approximately 280 acres of ponds where the slurry would evaporate. The dried sludge cake could then be trucked to a nearby landfill or made available to the public for use as fertilizer and/or soil conditioner. Edwards Air Force Base, however, showed no inclination to support the project.

Meanwhile, perpetual candidate and incumbent Mayor Sam Yorty was facing his final political campaign in 1973. Yorty attempted to replay the 1969 campaign, in which he defeated Councilman Tom Bradley with charges that Bradley was a dupe of left-wingers, would bring about a black takeover of city offices and was a friend of the Black Panther Party. But memories of the 1965 racial violence in Watts, which Yorty had previously successfully used, had faded in a city known for continually reinventing itself. Los Angeles was now enjoying civic quiet and general prosperity. Bradley came out swinging following a successful April primary win. In frequent community and television appearances, the tall, twenty-one-year veteran of the Los Angeles Police Department conveyed quiet strength and determination, and displayed a contained charisma and presence as he counterattacked Yorty at every opportunity. Bradley won the election with a large majority of the votes.

Bradley immediately changed the composition of the Board of Public Works and appointed the first woman, the first African American and the first Mexican American to the board. Since Bradley's inaugurating actions, there has been minority representation on the board. Moreover, Bradley began a tradition of appointing environmentalists and community activists, altering the previous all-white, all-male, industry- and business-dominated boards. By 2008 the Board of Public Works was made up of four women and one man, and included a community activist and an environmentalist. Moreover, Bureaus in the Department of Public Works were headed by women, African Americans and Asians.

But, at first, Bradley was not an environmentalist when it came to the sewers. Like Yorty, he supported Donald Tillman and other city engineering staff who assured him that the city did not need to stop its practice of pumping sludge into the ocean. He accepted the assertion that full secondary treatment for the sewage at Hyperion was unnecessary.

Bradley supported city efforts to convince federal and state officials that the sludge was not hurting the marine environment. Sanitation officials were certain that a soon-to-be-issued, three-year study by SCCWRP would prove that the water near the city's two ocean outfalls was clean enough. The study, released in September 1973, essentially supported Tillman's claim that sludge was not harming the marine environment, although it did point out the terrible effect of earlier DDT and PCB deposits on sea life.

The 531-page report of a three-year study of the Southern California Bight—from Point Conception to the Mexican Border—reported that there was no need for any "substantial modification of current wastewater disposal practices," and enthused about the lack of heavy metal presence in fish near the outfalls. The report also suggested that the city improve its primary treatment to "remove more floatable materials," because of the presence of discoloration in the water. The SCCWRP report, however, did not allay the concerns of the federal government, which then sued the city to force it to comply with "sludge-out" requirements.

Because of the federal suit, in November 1974 the counties of Los Angeles and Orange joined with the City of Los Angeles in an area-wide joint powers agreement (LA/OMA) to study how best to eliminate sludge from ocean discharges. While LA/OMA was evaluating alternative solutions for disposal of sludge, in December state and federal regulatory agencies demanded that Los

Angeles submit a facilities plan for its wastewater by 1977. The EPA and the State Water Resources Control Board refused to release funds to build the Sepulveda Basin water reclamation plant until the facilities plan was completed, and then the EPA sued the city to force it to complete the facilities plan. Tillman said that the EPA took the city to court, "not because we were polluting. The reason was unbelievable to me. City Hall, the council, couldn't convince the mayor or the city attorney to fight the EPA over this. They took us to court because we were not making satisfactory progress on the facilities plan." Although he admitted putting the facilities plan "on a side burner," he was certain they could "fast track it when we got a little closer" to the deadline. They were ready to build the Sepulveda Plant. "I spent an awful lot of time lobbying here and there to try to get us a release to build it," he recalled. As if he were re-engaging in the argument more than twenty years earlier, he said, "Let the money flow. The money was there." But the federal and state money did not flow.

In August 1975 the State Water Resources Control Board issued an NPDES permit that required the city, in a three- stage program, to cease ocean discharge of sludge by April 1, 1978. Meanwhile, in an effort to raise money for "federally mandated programs for the acquisition, construction, improvement and financing of sewage and wastewater collection, treatment, disposal and reclamation facilities," the City Council approved a bond measure for ninety-six million dollars, to be funded by a sewer service charge, for the November 1976 election.

Voters defeated the measure, in part because of the growing anti-tax sentiment throughout California, but certainly in large part because many of the city's leaders did not think that improvements to the system were necessary. Despite Mayor Bradley's and Los Angeles Times support for the measure, the City Council had publicly denounced the new requirements for sewage treatment.

With the defeat of the revenue measure, council members Yaroslavsky and Nowell led a battle to defeat the EPA sludge-out requirement. In December, the City Council voted to approve Tillman's interim sludge-out proposal. Yaroslavsky and Art Snyder (Nowell was absent) walked out of council chambers. A week later Yaroslavsky threw down the gauntlet in the L.A. Times editorial pages. Arguing that the "best scientific opinion" finds the ocean disposal the best system possible, Yaroslavsky wrote that the sludge "will have to be transported across miles of surface streets and freeways before it can be disposed of. The hazards inherent in such a system are self-evident. Moreover, the liquid, bacteria and toxic waste content of the sludge will render the landfills useless for decades to come." Confident that the city could win in a legal battle against the EPA, Yaroslavsky concluded that "Los Angeles should not spend one dime on EPA's mad scheme until the environmental issues are settled in court."

Enter the Pacific Legal Foundation. The forerunner of many similar conservative, public-interest, nonprofit organizations, since 1973 it has represented a cross-section of topics ranging from property rights to the constitutionality of affirmative action. Pacific Legal also fought land use restrictions by the California Coastal Commission and waged battle on rent control statutes throughout the United States.

On February 15, 1977, council members Yaroslavsky, Nowell, Snyder and John Ferraro, joined by the Pacific Legal Foundation, sued the federal government to stop the interim sludge disposal project. Pacific Legal and the council members argued that the EPA had not prepared an environmental impact study on the interim project or on its decision to eliminate all ocean disposal of sewage sludge, and that the EPA failed to take into consideration ocean disposal as an alternative. The foundation would later file several additional suits on behalf of the city. Although one suit did eventually reach the U.S. Supreme Court, none were successful in the end.

Six months after the Pacific Legal suit was filed, the EPA sued the City of Los Angeles. The EPA claimed that the city was violating the NPDES permit granted in 1975. The city had not met several important deadlines, including submitting plans for discontinuing the discharge of sludge and submitting interim reports on upgrading to full secondary treatment. It had also failed to submit a facility plan originally due in 1975. The EPA asked the district court to impose a schedule of completion for modifications at the plant and to enjoin the city from any future violations of the permit.

Tillman was convinced that the EPA was using the city suit to achieve nationwide sludge-out and national full secondary treatment. But, in fact, no other city had at that point refused to stop discharging sludge into the ocean. While motions were filed and judgments were given on several of the Pacific Legal suits, the engineers continued work on the citywide wastewater facility plan. This plan called for dewatering the sludge with solid bowl centrifuges and then hauling the wet cake product to a sanitary landfill for burial. Because of the ongoing LA/OMA study, the plan did not include any specific sludge-out program.

In April 1977, LA/OMA submitted a report that recommended building a new, state-of-the-art, multi-million-dollar sludge burning system, the Hyperion Energy Recovery System (HERS). HERS would burn the sludge and produce energy in the process. First, new centrifuges to reduce the amount of water in the sludge coming from the digesters would be installed. The sludge would then be carried to a sludge-drying facility that featured a unique, technologically advanced, patented system (the Carver-Greenfield Process) that would fully remove all liquid.

The process, previously used for animal rendering and to handle waste at a large brewery, had never been used to dry sludge. HERS included a vacuum and heat exchanging drying system that relied on a special oil to give the drying sludge mobility. The sludge would be dried to a fine powder. The powder would then be burned in huge reactors that would create steam to run two turbines. The ash would then be hauled offsite to a landfill. Gas drawn from the sludge digesters would be used to run four gas turbines that would join the steam turbines to produce power for the plant and replace the polluting diesel engines then running on digester gas.

By late 1979, Mayor Bradley was ready to reach an agreement with the federal and state regulatory agencies regarding sludge disposal. Ray Remy, Bradley's chief of staff since April 1976, was known as a skillful negotiator and began working with state and federal officials to resolve the sludge disposal problem. In 1994 he recalled the discussions that in 1980 led the city to sign a consent decree to stop discharging sludge. "We were reaching the point where we might not be able to have any additional building permits because of the lack of capacity for the San Fernando Valley—and Hyperion still needed a major upgrade," he said. Because there was some urgency to reach a settlement so that the city could begin work on the Sepulveda Basin Water Reclamation Plant, in June 1980 the city signed a consent decree to stop discharging sludge.

The consent decree called for sludge out of the ocean no later than July 1, 1985 and established a schedule for the city to complete construction on HERS. In addition to a $2.6 million fine for its previous violations, the city agreed to establish an Environmental Trust Fund for projects to improve the environment. Work on the HERS project began in 1981.

Meanwhile, El Segundo, which had for years been suffering from odors from the plant, attempted in vain to halt the HERS project. Officials at El Segundo were not the only critics of the proposed system. Board of Public Works Commissioner Royal O. Schwendinger, who would be in charge of overseeing construction of the project, believed the operating expenses for the program would dramatically exceed any financial benefits from selling excess electricity generated. Bureau of Sanitation Assistant Director William Garber was also skeptical about the benefits, and argued that the nutrients in the sludge were actually beneficial to sea life.

George O'Hara, chief engineer at Hyperion, was afraid that he might be left with "an unmanageable white elephant." But Roger T. (Tim) Haug, project leader for HERS and chief technical consultant to the city, dismissed such concerns. In 1984 Haug insisted that "state-of-the-art gas and steam turbines and sophisticated air pollution controls would dramatically reduce daily emissions by nearly three thousand pounds." The primary attraction of the project, Haug said, "lies in its energy efficiency. Because the new system will rely on high speed rather than high heat, little energy is expended in the drying process. The energy saved in drying plus that recovered in burning is at the heart of what will make the Carver-Greenfield Process an energy producer rather than an energy user." At the time, HERS appeared to be the best possible solution for the city's sludge problem. And though the total project cost was estimated at about $185 million, per the Clean Water Act provisions the city would only pay two and one-half percent of costs.

In December 1977 Congress approved a new amendment to the Clean Water Act of 1972 that raised the hopes of city officials. The amendment allowed cities with deep ocean discharges to apply for a waiver from full secondary treatment. The 1977 amendment appeared to be the solution to the city's intractable refusal to go to full secondary treatment of its sewage. In September 1979 the city filed the application for the waiver to Section 301(h) of the Clean Water Act. In November 1981, EPA administrator Paul DeFalco issued tentative approval for the waiver from full secondary treatment. A hearing to determine if the waiver would be permanently granted was scheduled for March 1985.

In 1982 Don Tillman retired. It would be three years before the Sepulveda Plant would be completed. A winner of many architectural awards, the plant was renamed the Donald C. Tillman Water Reclamation Plant in honor of the man who devoted so much time and energy to its design and construction. Set on a ninety-acre site leased to the city by the U.S. Army Corps of Engineers, the plant initially treated 40 MGD of sewage. It included a six-and-one-half-acre Japanese garden built around a series of interconnected lakes that featured imported stone lanterns, an elaborate waterfall, a variety of Japanese bridges and an authentic Shoin building. The new sewage plant was so beautiful that weddings and special events were held there almost from the first day of operations. *Los Angeles Times* architecture critic Sam Hall Kaplan noted that the plant, designed by Anthony Lumsden, was "one of the most incongruent architectural projects in Southern California and one of the region's oddest attractions." The site of dozens of motion pictures and hundreds of commercial photographs, the plant has attracted thousands of visitors to its parklike grounds and architecturally stunning administration building since opening to the public in 1985.

Federal and state authorities had approved financing for a Los Angeles-Glendale water reclamation plant in 1973, because the size of the plant would not encourage growth as they had assumed the Sepulveda Basin Plant would. Unlike the Sepulveda Plant, which was designed to eventually treat 200 MGD of sewage, the Los Angeles-Glendale plant was planned to treat 20 MGD. It would be located next to the Los Angeles River, on land owned by the City of Glendale in an industrial area in Los Angeles. It opened on May 25, 1976 and began discharging secondary-treated effluent into the Los Angeles River, finally allowing the city to close the Valley Settling Basin built in the 1950s.

In the late 1970s and 1980s, despite an aggressive campaign by the city's Board of Public Works to reduce industrial pollutants coming into the city sewers—including shutting down several plating companies and sealing their sewers with cement—the Regional Water Quality Board reported that "the plant's wastewater discharge exceeded daily discharge standards for arsenic on ten occasions, cadmium on twelve, mercury on twelve, ammonia on twelve and cyanide twice." What is more, the city was continually violating effluent standards for biological oxygen demand and release of oil and grease.

Noncompliance and corrective action reports filed with the Regional Water Quality Control Board between 1979 and 1984 documented an increasing list of violations, due to a lack of maintenance at the plant. While city engineers were concentrating on HERS and other planned future improvements at Hyperion, the city's own annual reports detailed the growing problems: "deteriorating condition of the primary sedimentation system arising from over thirty-year service life of the facility while two or more primary tanks were often out of service." Electrical outages and occasional staff indifference caused chlorinated secondary effluent to be diverted through the 1-Mile Outfall.

During this time, plant manager George O'Hara recalled, "We had a series of challenges." Occasionally a digester would be taken out of service to clear out decades of accumulated grit and debris that had decreased its capacity. O'Hara added, "It would then take nine months to clean a digester and you would run into air quality problems." When sludge was emptied out of the digester, the muck would be held in a pond, until it dried, at the northeast corner of the plant, before being trucked to a landfill. The muck, in O'Hara's understated words, "caused a potential for odors." These foul odors then settled in reeking vapors over neighboring El Segundo. O'Hara said the area was the size of half a football field. The corner of the plant faced the dead-end of Pershing Drive, a wide road west of the Los Angeles Airport. Sometimes a drunk driver—but often a driver simply unfamiliar with the wire fence enclosing that area of the plant—would speed down Pershing, crash through the fence and find himself mired in sludge, only to be rescued by the El Segundo Fire Department.

"Hyperion was one of the largest activated sludge sewage plants in the United States," Plant Manager O'Hara asserted, "and maintenance was not a high priority. We were always begging for money." Requests for money needed approval by the city administrative officer, who approved all city departments' requests for funds above their budgeted allowances. O'Hara said, "their job was saving money and they did a good job." Unfortunately, O'Hara's boss was equally uninterested in spending money. "The answer from top management," he recalled, was, "If it doesn't cost money, then fix it." Chewing gum, rubber bands and paper clips didn't quite do the job.

Jack Betz, sanitation bureau director, had a reputation for being something of a tightwad. When questioned about that reputation, Betz, who retired in 1981, acknowledged, "I was always interested in saving as much money as possible and operating as efficiently as possible." Had the Bureau of Engineering, responsible for sewage design and construction, requested the funds, Bureau of

Sanitation personnel believed the CAO would have approved the requests.

Although Betz said it was a "give-and-take" situation, the engineers at the Bureau of Engineering regarded the engineers at the Bureau of Sanitation as, at best, second-class. Engineering rarely supported requests from the Hyperion plant. As one deputy engineer said, "engineers design and build things. They don't operate and maintain sewage plants and pipes." Moreover, during this time, Tillman was leading wastewater battles on several other fronts. The Bureau of Sanitation was the city's stepchild during a period of deteriorating conditions at Hyperion and the North Outfall.

Because the city had not yet built the new plant in the San Fernando Valley, many more millions of gallons of sewage than the pipe was designed to carry were pouring through the North Outfall. The overflow problems were also due to a deteriorating North Outfall Sewer. A major overhaul of the aging sewer was necessary. It would be several years before a new interceptor would be built to divert sewage from the North Outfall while repairs were made on it.

In 1985, just as city officials were expecting to receive final approval for the waiver from full secondary treatment, the overflows at the plant and at Ballona Creek became the city's Waterloo. Although the Hyperion plant had experienced more than thirty "emergencies" in 1984, during which time city engineers had frequently diverted chlorinated effluent to the 1-Mile Outfall, the city expected continued leniency from the state and the federal government. To begin with, HERS was being built, and it would eliminate sludge discharge to the ocean and supposedly greatly reduce the violations of the past. Also, the water reclamation plant in the San Fernando Valley was expected to go online within a few weeks of the meeting. When it did, this plant in the Sepulveda Basin would relieve, by 40 MGD, the burden of treating sewage at the Hyperion plant, then treating in excess of 400 MGD.

A pro-forma hearing to approve the city's request for a 301(h) waiver was held on March 25, 1985. The EPA's DeFalco had tentatively approved the city's request four years earlier. The hearing was attended by a dozen city officials and by Dr. Rimmon Fay, the single opponent to the city's waiver request. Dr. Fay, a marine biologist and former member of the California Coastal Commission, attempted to convince the regional authorities to withhold the waiver. Fay, who headed a biological marine supply company in Venice and regularly trawled the waters of the bay, insisted that marine life had declined across the board in recent years, so the city should not get the waiver. The board listened to him but simply stated that it would have a final decision that summer. The bureaucrats went home happy; Rim Fay left dispirited.

Fortunately, a few days later, as he returned from one of his trawling expeditions, Fay met frequent beachgoer and long-distance swimmer Howard Bennett, a Culver High School teacher who, with a flair for the hyperbolized phrase, would soon jump into the spotlight of celebrity as he took up the cause of environmental protection of Santa Monica Bay.

GALLERY

Decades of Progress
1946-2008

Mayor Fletcher Bowron and engineers examine fertilizer from sewage. circa 1952

Repairing 1907 Central Outfall Sewer. circa 1948

Interior of 1907 Central Outfall Sewer. circa 1949

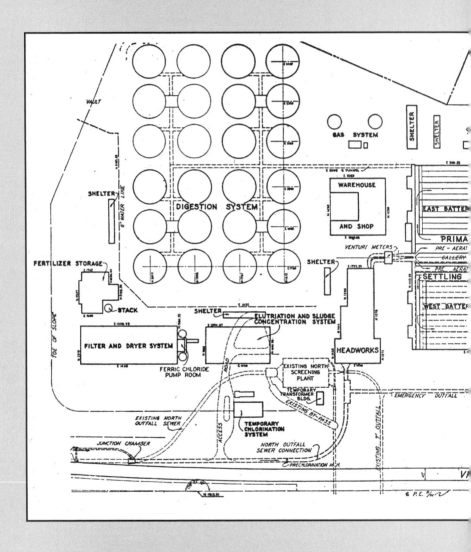

Schematic drawing for 1950 Hyperion Treatment Plant.

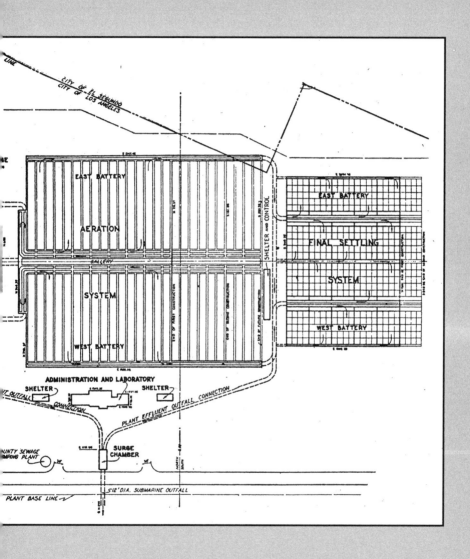

Construction of sludge digesters. 1950

Mayor Fletcher Bowron christens primary tank at Hyperion plant. 1950

Hyperion aeration tanks on a clear day in 1953.

A Hyperion Treatment Plant worker attempts to hose off detergent suds that continually escaped from aeration tanks, plaguing the plant and the adjacent city of El Segundo throughout the 1950s and 1960s.

Northwest view of Hyperion Treatment Plant with a compressor tank in foreground. 1953

Sunbathing at the beach near Hyperion in the 1950s.

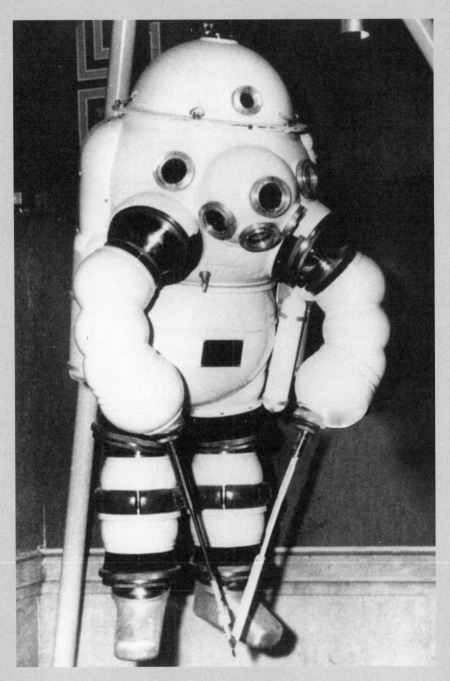

A diver's suit was necessary when inspecting the ocean outfall in the 1950s.

Low tide exposes the 1-Mile Outfall.

A sign warns of the danger of sewage contamination. circa 1955

Signs like this, posted by health department, warn of the dangers of swimming in the sewage-contaminated Los Angeles River. Despite warnings, children continued to play in the riverbed. 1955

Construction began on the 7-Mile sludge outfall from Hyperion into Santa Monica Bay. 1956

Contractors built two new ocean outfalls to Santa Monica Bay in 1956, a 5-Mile Outfall for discharge of treated sewage and a 7-Mile Outfall to carry sludge to a 350-foot canyon.

always-in all ways **DEPENDABLE**

CLEVELANDS on the oil trenching jobs—main lines, distribution lines, service lines, gathering lines, stripping pipe—can be depended on to deliver maximum profitable performance. The proof is in the records. CLEVELANDS have the stamina, power and strength to dig in and lick the toughest going over the most rugged terrain and through the meanest soils. They have the exceptional maneuverability to assure ease of operation in the tightest places and a wide range of transmission-controlled speed combinations giving under all conditions the best and fastest speed for the work at hand. Add to this low fuel consumption and minimum maintenance and you'll know why oil field owners depend on and swear by their CLEVELANDS.

THE CLEVELAND TRENCHER CO.
20100 St. Clair Ave., Cleveland 17, Ohio

◇   SEE YOUR LOCAL DISTRIBUTOR   ◇

SMITH·BOOTH USHER CO.
2001 SANTA FE AVENUE
LOS ANGELES 54, CALIFORNIA

SHRIVER MACHINERY CO.
1756 GRAND AVENUE
PHOENIX, ARIZONA

A tongue-in-cheek advertisement promotes trenching equipment for sewers. 1956

The Los Angeles-Glendale Water Reclamation Plant was established adjacent to the Los Angeles River in 1976.

The city used boats to sample Santa Monica waters for sewage contamination. The vessels ranged from a 1950s converted trawler to a yacht completed in 1990. The trawler, the *Prowler*, was purchased in the late 1950s and was equipped with biology laboratory equipment. The *Marine Surveyor* was specifically built for the city and replaced the trawler in 1964. Originally bid by a contractor at $441,000, a new oceanographic monitoring vessel, *La Mer*, was a yacht designer's idea of a floating laboratory and cost the city more than six million dollars by the time it was completed.

The Central Outfall Sewer, built in 1907, collapsed in 1989 at the entrance to Hyperion Treatment Plant, and it was quickly repaired by crews working around the clock.

Top: The D.C. Tillman Water Reclamation Plant treats eighty million gallons of sewage a day and began operations in 1986.

Right: Dorothy Green, founder of Heal the Bay.

Sampling water in Ballona Creek.

Terminal Island Treatment Plant.

Full secondary-sewage-treatment egg-shaped digesters at Hyperion. 1988

To enrich the soil, approximately seven hundred tons of sludge (biosolids) are applied daily at the city's 4,600-acre Green Acres Farm in Kern County, which was purchased in 2000.

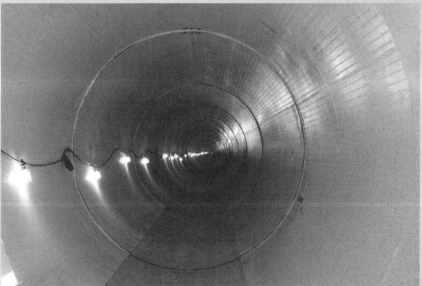

The East Central Interceptor is eleven feet in diameter, eleven miles long, flows an average seventy feet underground, and carries more than 230 million gallons of sewage a day. 2004

Top: The East Central Interceptor Sewer was built in 2004. The reinforced concrete base sewer is fitted with a plastic form.

CHAPTER TWELVE

Do-Gooders Get Organized

Howard Bennett told a *Los Angeles Times* reporter that he had been awakened to dangers lurking in ocean waters when he was about to enter the water but was "stopped by a fisherman," who "was waving his arms and saying the water's poisoned." The "fisherman" was Dr. Rimmon Fay, a serious oceanographer concerned about pollution, who probably didn't use quite those words. The marine biologist was intense, but not simple. Fay had always spoken and written in more measured terms about the city's waste disposal that he believed "degraded the environment" and was "harmful to fish and aquatic organisms sensitive to toxins and excessive organic matter." Still, Bennett, a long-distance swimmer for more than twenty years, got the message and became a Cassandra for a new, organized local environmental community.

Bennett contacted Dorothy Green, who was then president of the Los

⊰ Rimmon Fay, Ph.D., inspired the founders of Heal the Bay.

Angeles chapter of the League of Conservation Voters and a veteran of numerous battles with government bureaucrats. In 1973 Green founded Californians for Campaign Reform, a group dedicated to public financing for political campaigns. She became the local representative for Common Cause and appeared regularly before various civic and political organizations. In subsequent years, as she continued to fight for public financing of political campaigns, she adopted the cause of preventing nuclear power plant siting by the city's Department of Water & Power. She went on to oppose the Metropolitan Water District's plan for the state water project, the Peripheral Canal. When the battle over the waiver of full secondary treatment developed, Green was a seasoned activist.

　　　She and Bennett were not the only environmentalists concerned about the waiver. Rim Fay also spoke to Long Beach environmentalist Don May, who talked to Felicia Marcus, then a lawyer with the Center for Law in the Public Interest. As an aide to Westside Congressman Anthony Bielenson, she had previously worked on environmental issues. Marcus also maintained a friendship with Judge Harry Pregerson, for whom she had clerked while in law school. Judge Pregerson oversaw several consent decrees affecting L.A.'s treatment of its sewage in the 1980s.

　　　Green began holding meetings at her home. She, Marcus and Bennett were soon joined by other activists, including Mark Pollack, Moe Stavnezer and longtime environmentalist Jamie Simons. For want of a better name, the group agreed to Howard's name for a new organization, the Coalition to Stop Dumping Sewage in the Ocean. They wrote to several environmental groups and a variety of newspapers, and appeared before the Santa Monica City Council, which a month later unanimously voted to oppose the waiver request.

　　　The environmentalists wrote the Regional Board demanding a second hearing. Other organizations, homeowners' associations, environmental organizations sent letters as well. The board responded to these demands and scheduled a second hearing for May 13.

　　　The opposition was never as passionate or well organized. City engineers, elected officials and appointed Board of Public Works members were depending on the expertise of their chief ally, Willard Bascom, director of the Southern California Coastal Water Research Project (SCCWRP), who had previously succeeded in convincing state authorities that the city's sewage treatment did not need to be upgraded. He and Dr. Fay had been at odds since as early as 1974, when Fay was representing the Ocean Fish Protective Association and Bascom was insisting that more research was needed before upgrading the sewer system. The dispute became especially heated in 1984, when Fay claimed that he "saw animals dying on the ocean floor, rotting in their shells." Bascom admitted, at the time, that a small percentage of the ocean floor was contaminated by DDT discharges from previous decades. However, he argued that "the fish population has been enhanced substantially by the food discharged into the bay in the form of sewage." He insisted that his studies showed "no damaging effects on marine animals," and swore that metals discharged with the sewage did not make any difference.

Except for a few charismatic and dynamic leaders, civil engineers are not attuned to political engagement. In the past City Engineer Robert Horii had dealt with city officials, and though among themselves he and his staff dismissed the environmentalists, they did not air their differences to the news media. Horii, who had only just that year been appointed city engineer, was the first Asian American to hold that office. To many, he appeared mild and self-effacing, but he was a seasoned civil engineer with decades of sewerage experience. He was a respected manager of a staff of several hundred engineers and would, in the future, gain the respect of environmentalists as he unemotionally mediated differences between his staff and the enraged activists.

Maureen Kindel was a different sort than previous Board of Public Works members. She was the first woman on the board and had no real business experience, but was a skilled fundraiser who, within two years of her appointment to the board, had led a successful battle to override a City Council effort to abolish the board. She had also supported a strong response to industrial pollution, creating what she called "The Toxic Cops," who went after firms violating their permits, sealing their sewers and recommending, successfully, that the city attorney prosecute these firms. Equally important, she was a confidante of Mayor Tom Bradley, who trusted her judgment.

The battle over the future of the city's sewerage practices would be waged with Kindel, city engineers and Willard Bascom on one side, with the environmentalists and Cliff Gladstein and Carol Kurtz, who worked for Assemblyman Tom Hayden, representing the "good guys." The struggle took shape on May 13, 1985, when the Regional Water Quality Control Board held a second hearing into the 301(h) waiver.

CHAPTER THIRTEEN

Healing the Bay

David Ferrell, a *Los Angeles Times* reporter, responded to the concerns raised by Howard Bennett and Dr. Fay. The day of the hearing on May 13, his three-page story about the waiver battle was on the front page of the newspaper. With a prominent photograph of Bennett at the beach in Playa del Rey, the story detailed some of Hyperion's recent state water quality violations, including exceeding daily discharges of ammonia, oil, grease and such heavy metals as nickel, zinc, cadmium, lead and chromium that were bonding in excessive amounts to the sludge and were flowing out the 7-Mile Outfall. Buoyed by the article, the critics showed up en masse for the May hearing.

Despite the willingness of the EPA and the Regional Water Quality Control Board to hold this second meeting to discuss and evaluate the city's

◁ Final clarifiers at Hyperion Treatment Plant. 1988

application for a waiver from full secondary treatment, city officials assumed the waiver was "a done deal." A staff report included in materials distributed at the hearing noted that the city's plan to improve primary and secondary treatment would "increase the volume of secondary-treated effluent from 100 to 240 MGD," and would therefore make a considerable difference in the city's effluent. Several city representatives, including City Council President Pat Russell, Maureen Kindel, City Engineer Horii, several members of his staff were at the hearing to defend the city's application.

The hearing, held in the dusty, cavernous assembly room of the old State Building on Broadway just a few blocks from City Hall, was noisy and unrestrained. The Coalition to Stop Dumping Sewage into the Ocean brought out an impressive five hundred people, including a busload of Bennett's students. The city was outgunned, outmanned and outmaneuvered. Opponents of the waiver were convinced that the city and the regional board had joined in a conspiracy to keep the ocean water dirty, and took to the microphone to say so. Individual after individual got up to complain about the terrible conditions in Santa Monica Bay. Bennett claimed that he had seen changes in the clarity of the water that "frightened" him. Students demanded to know why the regional board was allowing their beaches to be polluted. Christine Reed, Santa Monica's mayor, voiced her opposition to the waiver. Dr. Rim Fay said, "you don't see large quantities of baby fish anymore. The beds of shellfish, the lobsters, the abalone that were present before, are no longer there." As the meeting dragged on, Dorothy Green had to leave, but Felicia Marcus read Dorothy's prepared remarks in a soft but impassioned voice. Representatives from several other environmental groups had their say. Former State Water Quality Board member Carla Baird reversed her earlier support for the waiver and told the Regional Board that "a great deal of evidence against the waiver" had appeared since 1977. The board turned to city representatives and asked them to respond to the questions raised by the public.

City Engineer Horii described the city's current plans to increase the treatment at Hyperion and bring it into compliance with state standards, which at the time were slightly less stringent than federal standards. The recently appointed Republican Chairman James Grossman, at first somewhat bemused by the commotion, listened to Councilwoman Russell, Maureen Kindel and Horii, but began to pick up steam as he asked a few sharp questions after they somewhat stoically presented their case. After approximately two hours, the members of the board agreed to take the waiver matter under advisement and promised a ruling a few months later. Despite the sometimes raucous hearing, the city engineer and elected officials were convinced they would receive the waiver.

The pressure continued to mount. On May 16, the Los Angeles *Herald Examiner* claimed that the city had violated its permit fifty times in the past year, editorialized against the waiver and promoted full secondary treatment, despite the additional cost. On the same day the *Los Angeles Times* noted that Carla Baird "didn't trust Los Angeles promises of improvement." Van Nuys State Senator Herschel Rosenthal and Westside Congressmen Mel Levine and Anthony Bielenson also publicly opposed the waiver.

Within days, other newspapers, the *Santa Monica Outlook*, the *Culver City News* and the *Wave* were reprinting accusations, including Bennett's ever-increasing and hyperbolic claims that he knew of "a boy who went surfing in '83 and came home with infected sinuses, had surgery to remove lesions and lost thirty pounds," and who, according to Bennett, was still under a doctor's care. The fifty-five-year-old Bennett exclaimed, "Over the years, I've seen changes in the clarity of the water that have frightened me. I can no longer see to the ocean bottom. I can't even see to the end of my arm now." It would have been surprising if Bennett had *ever* been able to see the end of his arm at the bottom of the ocean, since currents and waves in the bay constantly churn up bottom sediment. But no matter. Bennett was reveling in the attention.

Shortly after the March hearing, westside State Assemblyman Tom Hayden charged the regional board with "an astonishing failure" to control toxic contamination in Santa Monica Bay. He was astonished to learn that seven hundred tons of DDT had been dumped into the bay through the Los Angeles County Sanitation District's sewage discharge at White Point near Palos Verdes. He scheduled an Assembly Task Force Meeting for the following May. In April, the California Department of Health Services issued a warning not to eat white croaker or Dover sole, because carcinogenic chromium and cadmium were concentrated in the livers of these fish.

Hayden's meeting was held on May 17, four days after the second board hearing on the waiver. Willard Bascom, head of SCCWRP, insisted there was little danger of sport fishermen getting cancer, but one of his key investigators, David Brown, head of the project's chemistry department, de-nied Bascom's reassurances. He testified that Bascom had "routinely withheld information that was damaging to sewage dischargers." He told the task force that Bascom warned his staff that they could be fired if they released "a study reporting high toxic levels in municipal mammals." According to Bascom, no such threat had been made. Furthermore, he argued, the sewage discharges, including the sludge discharge, were good for marine life.

Especially damaging was Brown's allegation that in a memo to the staff (portions of which were reprinted in the *Times*), Bascom reminded them that "cities and counties of Southern California are willing to pay for our studies, because they are engaged in a long-term struggle with the EPA over whether secondary treatment is required." Brown responded with an angry letter accus-ing Bascom of providing false information to the public.

On Sunday, June 2, the Coalition to Stop Sewage Dumping marshaled a hundred people, who marched around City Hall, waving placards. They un-raveled a wide, mile-long brown ribbon, festooned with petitions. The outsized ribbon, the outlandish claims of dead jellyfish, dead barnacles and the absence of sand crabs on the beach combined with the presence of people dressed in ocean-themed costumes, became a magnet for television and newspaper report-ers. A few days later the cities of Beverly Hills and West Hollywood joined Santa Monica in calling for full secondary treatment. Willard Bascom wrote an opinion piece in the *Times* that insisted the bay was on the mend and that sludge would add to air pollution if burned.

The city might have withstood the public outcry and received the waiv-er if the sewer system had not sprung a leak. On July 12, 1985, several weeks after the May hearing, raw sewage poured into Ballona Creek. It would be the first of many spills. Although relatively small—three thousand gallons or so, as the city initially claimed—the spill went unreported. Instead, two county health officers discovered the sewage spill from the overflow structure—a con-crete box at Jackson Avenue that had been installed in 1964 for storm water overflows—while they were taking water samples that morning.

Unfortunately for the city, Dorothy Green's brother was working on a construction project near the overflow on the same day. "Well," she said, "my brother got splashed. He came home and asked me about what was going on over there." She saw an opportunity to "embarrass the city," called Tom

Hayden and the television news stations and "we held a press conference right by that concrete box." The city engineers were nonplused when three more spills happened in July. They knew that wet weather could result in overloading the North Outfall, but in their minds, summer flows should not have overtaxed the pipeline. As newspapers described the "Eco-Mess" afflicting the bay, everyone was claiming that the waiver could be a death sentence for the sea.

A year before the hearing on the waiver, Delwin A. Biagi was appointed Bureau of Sanitation Director. Maureen Kindel had grown increasingly concerned about lack of maintenance at the plant, growing reports about employee dissatisfaction and what she saw as a "business as usual" approach by bureau management regarding other sewer issues. She had asked Biagi, who had little experience with the sewer system, to leave the Bureau of Engineering, where he was a deputy city engineer, to take on the job at Sanitation when Bureau Director Tad Isomoto retired. Biagi accepted the job after Kindel assured him he would have the full support of the board.

"The second day I was here," Biagi recalled, "the effluent pipe went down. Folklore has it that because Tony, the chief electrician, wasn't here and didn't use his hair dryer to dry off the points on the motor control units, we blew a bunch of fuses." When the fuses blew, the power to the pumping plants went out and, said Biagi, "we began discharging primary effluent out the 1-Mile Outfall, and Garber gave me a three-page note explaining it all: 'We don't have to tell anyone yet. We don't tell anyone till we know when it's fixed.'" Biagi added that a few days later he called Ghirelli, who suggested that "I let them know a little earlier than Garber wanted." In those days, Biagi said, "the relationship with the regional board was that we would give them all these excuses why we couldn't meet our permit limits and they would accept them all and everybody would go on their way." This amicable relationship took a turn in the summer of 1985, when legislators, environmentalists and the press began excoriating the city for its sewage mess.

On July 20, more than thirty thousand gallons rushed into Ballona Creek. On July 22, a spill lasting a little more than five minutes flowed into Ballona. On August 9, Ghirelli recommended that the board fine the city $30,050 for repeatedly allowing raw sewage to overflow into Ballona Creek. On August 21, the city agreed to pay the fine, after deciding to waive the hearing later that week. In addition, the city planned to install an automatic chlorination system within the next two months. That would make unnecessary for city employees to have to shovel chlorine salts into the sewage whenever it overflowed into the creek.

As local newspapers began running lengthy three- and four-part series on the spills, Kindel scheduled a series of meetings with the editorial boards of both the *Los Angeles Times* and the *Herald Examiner*. It was to no avail. While television reporters denounced city officials, the newspapers editorialized that the waiver would be the sea's death sentence and the *Santa Monica Evening Outlook*, in endorsing refusal of the waiver, described Santa Monica Bay as "our toxic coast."

At the same time as the spills were occurring, construction on the Hyperion Energy Recovery System (HERS) was adding to the bad quality of the effluent. Construction crews frequently disrupted routine maintenance and operations. Several digesters, the sludge containers that anaerobically reduced the volume of the sludge, were taken out of service for repair and cleaning. Unfortunately, delays in construction meant that the city was not going to the meet the July 1985 deadline for completion of HERS originally agreed upon in the 1980 consent decree.

The city had signed contracts for construction of HERS in the summer of 1983. Harry Sizemore, a senior engineer, said there wasn't enough time to meet the consent decree deadline for summer 1985. "With completion deadlines for early spring to summer 1985, we knew we were going to miss the deadline," he explained. Several of the components had been tried in animal rendering and beer processing plants, "but these processes had never been implemented anywhere in the world before in a sewage treatment plant," he recalled. "There was an awful lot of new technology. The extraordinarily rapid timeline was very optimistic. It was mandated by the EPA for political reasons, nothing technical, nothing environmental. They just wanted to get it done by 1985. Any delays would make it impossible." And there were many delays.

But the city met regularly with officials from both the California and the federal EPA and, said Sizemore, they felt it was making reasonable further progress and therefore the time delays didn't seem a concern. In the spring of 1985, before the sewage spills, city officials gave Judge Harry Pregerson a revised schedule for completion of HERS. On August 27, the judge granted the extension, but only to February 15, 1986. Six months was not nearly enough time to meet the deadline for completing HERS and getting sludge out of the ocean.

The extension angered the environmentalists, by then seeking a role in the consent-decree process. The amount of solids leaving the plant increased. Sanitary engineers like to describe sewage treatment as "nature's way" of handling the waste. During that summer and fall of 1985, Mother Nature had become like Pliny the Elder's merciless stepmother, giving the engineers more trouble than they could bear.

During this time Kindel opened a door to the environmentalists. As she and other officials were meeting in a room near the board offices—a meeting that Kindel had called to bring together representatives from state, county and various city agencies to discuss and find solutions to the ongoing sewage crisis—Dorothy Green appeared at the meeting-room door and demanded that the meeting be opened to the public. Although the Board of Public Works had not responded to the environmentalists, Kindel acknowledged that Green was sincere. She invited her to join them and offered Green lunch. Although the bureaucrats from several agencies were unhappy with Green's presence, Kindel turned the corner in the debate about secondary treatment.

As Kindel later recalled, she had been conferring with Bradley constantly. "I would talk to him every day." What's more, she added, "I could also talk to Harry Pregerson, because my husband was a judge." Her husband, Stephen Reinhardt, and Harry were colleagues, both judges on the Ninth Circuit Court of Appeals. While listening to the city engineers, she felt increasingly uncomfortable with her limited knowledge about the system and its history. So, she asked Bradley's legal counsel, Mark Fabiani, to go through the boxes of detailed information piled in her office. Fabiani had been an intern with Reinhardt after finishing law school, and at Kindel's urging joined the Bradley administration in 1985.

Kindel asked Fabiani to go through boxes and boxes of information on sludge-out and full secondary treatment and told him, "I want you to give me an executive summary." Since Fabiani also knew Felicia Marcus and Harry Pregerson, Kindel added, "it was all sort of sub rosa. But it had to be sub rosa, because of the litigation. Sometimes you have to have informal channels. Three days later," Kindel added, "the boxes were all back and a thick report was sitting on my desk."

Fabiani's report laid out the major milestones in the system over the past twenty years, identified the immediate problems and solutions, outlined the necessary financing for both full secondary and improved partial secondary treatment, compared Los Angeles to five other major cities and to the L.A. county system, and outlined a $1.3 billion program for improvements. Kindel recalled that Fabiani said very definitely that "the city was in some peril if we followed the advice of the engineers at this time. He made it very clear to me. Not a fiat." As the political fallout from the spills continued, on September 4 Bradley announced that the city was now committed to installing a full secondary treatment facility at the plant. He did not, however, withdraw the city's request for the waiver.

Unfortunately, the North Outfall continued to plague the city. On Saturday, September 21, one hundred thousand gallons of raw sewage poured into Ballona Creek. Although regulatory agencies were notified, neither the mayor nor the public was informed until several days later. Environmentalists had now created a new organization, Heal the Bay, with a name they believed connoted hope instead of despair about the past. Despite Kindel's earlier overture, Heal the Bay was dismayed by the September spill, and many supporters became convinced that the city could not be trusted and would never complete the needed improvements in the system.

The consequences of the September spill would linger for more than a decade. Mea culpas were heard from everyone. An extensive public reporting system was devised. Kindel insisted on posting beaches immediately after learning of a spill, instead of waiting for the Health Department to take its authorized responsibility—usually at least two days after an incident. The political fallout was even greater. Gearing up for the 1986 gubernatorial campaign against Bradley, Governor Deukmejian attacked Bradley as a major polluter. Once seen as an environmental candidate, Bradley, who had been attacking the governor for permitting toxic waste sites, was now rapidly losing support of the organized environmental community. In an effort to stem the sewage tide, Bradley ordered the construction of two huge tanks whose combined capacity was one million gallons of raw sewage; the tanks would be built adjacent to the North Outfall overflow structure. When they were completed, the complex would be called the North Outfall Treatment Facility. Ironically, the treatment consisted of nothing more than what had been done in the 1920s: settling and chlorination.

Sanitation personnel were convinced that the spills were primarily due to the delayed opening of the Tillman Plant in the San Fernando Valley. Too much water was flowing into the North Outfall, and the Tillman Plant would initially divert at least 20 MGD from flowing to Hyperion. The plant had been scheduled to start treating the Valley's sewage in May 1985, but the chlorine used to disinfect the final effluent killed the fish in the river, so the plant opening was delayed until a dechlorination facility could be completed. The delays caused overflows at Jackson Avenue and a resulting public-relations nightmare that energized the environmental community.

At ten o'clock in the morning of September 25, 1985, the Donald C. Tillman Water Reclamation Plant began operation. The treated water was released to the Los Angeles River. An October 1985 joint report from the Bureau of Engineering and the Bureau of Sanitation stated that despite the relief from the Tillman Plant, flow to Hyperion was still exceeding 400 MGD, creating additional problems for treatment.

Although spills at Jackson Avenue were now limited to overflow during heavy rainfall, the Hyperion plant was stressed and the treated sewage often did not meet the city's permit requirements. Moreover, the Board of Public Works was increasingly concerned by SCCWRP director Bascom's reports. Despite years of sampling and testing by the city and SCCWRP, there was disagreement among members of the scientific and engineering community about the extent and nature of the impact of Hyperion on the bay. Bascom's reports, they noted, were complex and often confusing. Within a year, Bascom retired and a new director, Jack Anderson, a Ph.D. scientist with impeccable credentials, was appointed and instituted a range of new procedures and scientific peer review for all future project reports.

As political observers began to comment on the damage the spills were doing to Bradley's image as an advocate for the environment, new sewage afflictions appeared on the horizon. Plant maintenance had deteriorated since the time in the mid-1970s when voters and elected officials refused to approve additional funds for upkeep or improvements. HERS construction was causing chaos as contractors would close roads without notice to maintenance personnel, and new projects would proceed as operators attempted to work around the problems.

The engineers in charge of construction were reporting to the Bureau of Engineering, while engineers working on plant operations were reporting to the Bureau of Sanitation. The two bureaus rarely coordinated work. Pump failure at the Hyperion plant in October 1985 caused one hundred thousand gallons of secondary-treated sewage to flow out the 1-Mile Outfall. On the following day, another pump failure resulted in 2.5 million gallons of sewage that bypassed secondary treatment and went out the 5-Mile line. Sanitation workers reported the discharges to the county health department and the regional board. However, they did not notify Delwin Biagi, their boss and director of the Bureau of Sanitation, or the mayor. Bradley was furious. In one of his rare displays of written temper, he wrote the board demanding an explanation about why he learned of the incidents from the news media. Then, when he confronted Biagi about the events, Biagi admitted he was unaware of them as well. "Such delays in notification are intolerable," Bradley wrote, and insisted that the board create a better system of notification. Biagi and Kindel developed a new information system that included a "Sani-Gram" that went from the bureau to Bradley, his chief administrative assistant, the deputy mayor and all board members whenever an untoward incident occurred. Two weeks later, the Regional Board levied a $150,000 fine against the city because of the July and August spills. Although the city could have appealed the fine, on November 21 it agreed to pay, foregoing what would have been costly litigation that, in the climate of the day, would probably not have succeeded.

In September, when Mayor Bradley announced his intent to go to full secondary treatment, he asked the Regional Board to upgrade the California requirements to full secondary treatment so that the city could continue to qualify for federal and state money, which it could no longer do under the existing standards. The state officials, now operating on a political agenda favorable to sitting Governor Deukmejian, did not respond to Bradley's request. On November 26, the Regional Board announced its new requirements for full secondary treatment and formally denied the city's request for the waiver. Several months later, the EPA concurred with the state's decision, and the city began preparing for the almost two-billion-dollar cost of upgrading the plant to full secondary treatment.

As conditions at Hyperion deteriorated, Kindel took action. On December 18, the Board of Public Works removed plant manager Kenneth Ludwig and Ludwig's supervisor George O'Hara, who was in charge of the other three treatment plants and whose office was at Hyperion. O'Hara went to the Terminal Island Plant and Ludwig was assigned to research duties at City Hall. The men were not demoted, having committed no real misdeeds. They were simply seen as ineffective leaders in a time when strong leadership was required. It was clear to the board that both men lacked the support of Hyperion operators and engineers.

Harry Sizemore, executive assistant to Biagi, was immediately given responsibility for plant operations. O'Hara remained bitter long after he retired, "I was a scapegoat. No one had given us the money we needed to improve conditions at the plant." Ludwig was more sanguine about the reassignment, but recalled his own trials and tribulations at the plant: "Like so many city programs the plant was just another citywide program that had to be balanced on the back of other programs, had to compete with other city programs."

Sanitation funds were controlled by the city administrative office (CAO), and had to go through a complicated budget process that meant dealing with often-unsympathetic analysts. How difficult was it to get money for repairs? "We couldn't get tank replacements," he replied. Ludwig was referring to the primary tanks—they needed to have at least twelve tanks in operation, but at best they had nine in operation. "Electrical conduits were grounding out, water was getting out, concrete was falling in from the walls, constantly jamming up the works." Ludwig remembered a time when he asked for money to replace the chains that pulled the skimming boards in the primary tanks. "They cut us down to no more than one thousand feet a year." So, they turned the chains instead of replacing them. What's more they were "constantly fighting the CAO's office, getting the feeling that you can't get anything. You see a lot of construction, new people being hired, new equipment." Everything was in a constant state of transition and, he concluded, "We weren't getting diddly."

On the same day that Ludwig and O'Hara were transferred, the board decided to seek outside help to run the plant. They asked Donald Smith of James Montgomery Consulting to evaluate present conditions and prepare an initial report within sixty days. At the time, Smith was supervising consultant for work on HERS. On February 21, 1986, Smith presented his findings: a two-inch document that laid out all of the problems and how to resolve them. He advised the city to modify the existing organization structure, which had multiple levels of supervision without interaction between the divisions, identified support resources necessary to plant operation and suggested strategies to deal with administrative problems. Regarding poor overall compliance of the NPDES permit, he wrote, "It is clear that the principal reason is limitations in opera-tion of primary sedimentation tanks." The digester cleaning facility and sludge treatment facility were not operating at maximum. The effluent pump stations needed an additional station and an extensive control system. Recognizing that construction of HERS had created a "two-plant" philosophy, Smith added, "The single most significant impediment to plant performance is the high vacancy rate." Approximately thirty percent of budgeted positions were vacant.

Several months later, the board decided to bring the consultant onto the city payroll as the only way to change conditions at the plant. Shortly after Smith arrived at the plant, he initiated a program of interim improvements. John Crosse, who later became plant manager, recalled that Smith "coalesced support downtown among the powers that be, and we were able to expedite some quick design and implementation of four or five projects that had a dra-matic impact on improving our effluent quality. They added chemicals to the settling process in the primary tanks, upgraded the aeration diffusers for sec-ondary treatment, doubled the process flow and improved the digester per-formance by upgrading the mixing and cleaning the tanks." Felicia Marcus, Crosse noted, "kept a lot of focus on the elected officials. And the purse strings loosened up. We got some quick money, a quick infusion of cash to do those interim improvements." No longer outsiders, Heal the Bay became partners in the city's effort to improve conditions at Hyperion.

The results of the interim improvements would be startling. In the summer of 1985, suspended solids that went to the ocean were at more than 120 milligrams per liter (mg/L). By the summer of 1986, suspended solids in plant effluent had fallen to 60 mg/L. By 1989 the plant was achieving 20 mg/L, well below the 30-mg/L limit established for full secondary standards. So, although the city would eventually build full secondary treatment facilities, the improvements launched by Don Smith improved the morale of Hyperion employees and, even more importantly, not only allowed Hyperion personnel to improve treatment of the sewage, but met full secondary standards for sus-pended solids going to the ocean. The city continued to meet the requirements for the next thirteen years until 1998, when all full secondary facilities were completed at the plant.

The whole issue of the previous decade could be traced to what Dorothy Green had seen in 1985: "Everything imaginable was wrong at Hyperion. Even the concrete sidewalks were rotted and falling apart. Morale was in the toilet." Several years later, reflecting on the city's problems at Hyperion, Don Smith explained that the application for the waiver put the city in a form of paralysis. "The city was waiting to find out if they had to go to full secondary. That was an excuse not to invest in the infrastructure. Clearly you didn't want to make a major capital investment if nine months from now you'd have to change. If the city had done things about Jackson Avenue and had produced the effluent that it is producing today, they could have saved the city one billion dollars in capital investment. You have to look at the political side of the issue. The voters spoke with Proposition 13 (the 1978 tax revolt measure): people don't want to spend money, so the council responded to that. The political people were running very scared and didn't want to spend the money. It was the easiest course of action at that point in time."

As the city moved to fulfill its commitment to full secondary and to improve conditions at Hyperion, Kindel brought people from the Department of Water and Power (DWP) into a meeting with Mayor Bradley and the city engineers. After they talked for hours, DWP representatives suddenly acknowledged, as Kindel paraphrased it, "a malfunction in one of its transformers that was causing these surges, and the surges were causing these overflows and they needed a certain amount of money to fix it—and so Bradley ordered it and the problem was fixed and the surges stopped." The disconnect between city agencies that caused these kinds of problems would not really change until more than two decades later, when in 2006 the DWP and the Bureau of Sanitation issued a joint, massive study of water reclamation possibilities.

A coalition of environmentalists petitioned Judge Pregerson, asking to be allowed to participate in future decisions regarding the city's sewer system. On May 29, 1986, Pregerson granted friend of the court (amicus curiae) status to these groups. "Their role," he wrote, "will consist of receipt of all past status reports, all current and future reports. They will have the ability to contact both plaintiff and defense attorneys to communicate ideas and comments and to receive updates on status of ongoing negotiations." Included in the order was the notation that Bob Horii was to give a tour of Hyperion to the amici. They would have the ability to object to any proposed settlement and to any modifications in the consent decree and they would participate in ongoing settlement negotiations. The all-inclusive aspects of Pregerson's action ensured the acquiescence of the city to the environmentalists' demands.

In October 1986 Los Angeles agreed to pay an EPA-imposed fine of $625,000 for violating federal Clean Water Act standards. It was the largest fine collected since the act was enacted in 1972. The city also agreed to a timetable for completing full secondary treatment facilities and eliminating sludge discharge to the ocean.

Due to delays in getting HERS operational, the deadline for sludge-out had come and gone. During a tumultuous meeting in 1986 with the EPA officials, attorneys for the Justice Department and the city attorneys, the EPA demanded an aggressive schedule, with milestones and exculpatory language. Don Smith told them the city wanted to set milestones that could be met. "I said we're not going to put ourselves in a position of failure, because if we have to, we'll go to court to assure that."

On February 12, 1987, after months of meetings with all parties—the environmentalists, the EPA, the regional board and the city—a final amended consent decree was filed with the district court. The decree set a deadline of December 31, 1987 for ceasing sludge dumping, full secondary treatment to be achieved by 1998 and for HERS to be fully operational by June 30, 1989.

By the time voters were asked to approve the first of three bond measures to correct the problems at Hyperion, construct new facilities there and build additional interceptor sewer lines, more than 250 stories about sewage spills, Hyperion spills and sewage-caused pollution of Santa Monica Bay had appeared in newspapers across the country including the *Chicago Tribune*, the *Boston Globe* and the *New York Times*. Though the turnout of voters on June 2, 1987 was perhaps the lowest in city history—12.86 percent—the majority overwhelmingly approved the five hundred million dollars in revenue bonds to finance improvements to the city's sewer and storm water system. Seventeen months later, on November 8, 1988, with seventy-two percent of registered voters casting ballots, almost seventy-five percent approved another $1.5 billion to continue these improvements. Four years after that, on November 3, 1992, residents supported continued renovation of the city's sewer system with approval for an additional two hundred million dollars.

Mayor Bradley's Chief of Staff Tom Houston resigned in June 1987 and was replaced by well-known environmentalist Mike Gage. In a meeting with Biagi, Sizemore, Smith and Kindel, that also included Dorothy Green and Felicia Marcus, Gage recalled being impressed by Marcus' questions about what was being done at the plant and why. "She did it very gently, but she clearly was calling into question a few things and making points with the guys who were 'well, er, uh, well, yes, there's this and there's that.' But when we walked out of that meeting I said, 'We've got to figure out how to get her working for us.'"

CHAPTER FOURTEEN

Brown Acres No More

In June 1989, Kathleen Brown left the board to seek office as State Treasurer. Mike Gage convinced Bradley to appoint Felicia Marcus, who remained on the board for several years. Previously, an engineer had always sat on the board. After Kindel, the composition of the board changed. No engineer has been appointed since. By 2007, four women, most with either community-activism background and/or environmental credentials, were Board of Public Works members. Bob Horii was the first Asian appointed city engineer. Others would follow. African Americans took their place on the Board of Public Works and were appointed as bureau heads throughout the department. The tradition of multicultural diversity in city government has continued through a succession of city mayors, through Mayor Antonio Villaraigosa, who was still in office when this book was written.

◁ **Interior of interceptor sewer. 2004**

Shortly after Gage took office, Tom Houston and former Deputy City Attorney Barry Groveman, representing a company called Applied Recovery Technology (ART), approached the mayor and the board with a plan to take half the city's sludge—approximately 750 tons a day—and ship it to Guatemala. Shipping the city's sewage sludge to Guatemala seems a bit unfriendly, to say the least. However, ART, a Virginia-based waste-recycling firm, had received EPA approval for its methods, which they were using elsewhere. The firm planned to ship the sludge to Puerto Quetzal, a harbor on the coast of the Pacific Ocean. The plan was to compost the sludge at the port site. The compost would be used for reforestation and agriculture. By this time, HERS was nowhere near completion and the city was looking for an interim solution to disposing of its sludge by the December 1987 deadline. After checking with the EPA, the State Department and the Guatemalan ambassador (who expressed approval for the plan, Deputy City Attorney Chris Westhoff said), there was no problem. Guatemalan Ambassador Oscar Padilla Vidaure met with Mayor Bradley and apparently assured him that his government would favor the proposal.

Bureau of Sanitation Executive Assistant Harry Sizemore initially lacked enthusiasm for what was an obviously dubious project. "We had a policy of not going to Indian reservations, so going to third-world countries wasn't an environmental dream." Nevertheless, he, Westhoff, Spanish-speaking Deputy City Engineer Ralph Valenzuela and board Commissioner Dennis Nishikawa went on what Westhoff described as "the trip from hell." After a bouncy two-hour bus ride from Guatemala City to Puerto Quetzal on unpaved country roads, they were greeted by armed men carrying machine guns. Sizemore recalled with a shiver that throughout the countryside people were walking around with machine guns, some mounted atop turrets on wooden towers. Guatemalan President Marco Vinicio Cerezo Arévalo had only been in office for one year, following decades of military coups and guerilla warfare. Board of Public Works members had been assured the president supported the plan. Although the city would be paying approximately six million dollars to ART, it was unclear how much money, if any, would be going to Guatemala. Sizemore noted that there were no sewage facilities in the entire country, with sewage going into pipes to lakes and to rivers, "where it was diverted to a free-flowing stream into the ocean." Although there appeared to be enough undeveloped land where the Guatemalans planned to use the sludge, Sizemore was leery. "It was extraordinarily hot and humid, it was one hundred percent humidity, and I wondered what would cause the water to evaporate out of the sludge. I was skeptical. I thought they would just dump it."

Despite Sizemore's misgivings, Westhoff and Groveman drafted a contract to present to the Board of Public Works. Public Works Commissioner Edward Avila expressed some concern about the plan. He was worried about the negative implications to the Hispanic community if the sludge went to Guatemala. Avila was reassured by Houston's law firm that "the Guatemalan Board of Health and the Board of Environmental Affairs are both supporting the sludge contract." A day before the board was to sign the contract, Westhoff was dumbfounded to learn that the ambassador was now saying that the deal was off. Guatemalan President Vinicio Cerezo had categorically refused to accept the proposal, despite earlier letters supporting the program. The ambassador claimed he had been duped.

Following that fiasco, the city issued requests for proposals to handle part of or all of the approximately eleven hundred tons a day of sludge the city needed to get rid of by the December 31 deadline. Chemfix, a company represented by former Bradley aide Fran Savage, submitted a proposal and in December won a contract to take the sludge and fix it chemically at a site adjacent to the Los Angeles Airport, where chemicals would be added to kill all pathogens, producing a clay-like material suitable for daily cover at landfills. In December, the city signed a contract with Chemfix, although it would be several months before the site was approved. Chemfix would take the city's sludge until January 1990, when the city cancelled the contract because of cost, odor and management problems. To the meet the December sludge-out deadline, on November 2, 1987 the city began trucking a daily load of approximately eleven hundred tons to the BKK landfill in West Covina and the Chiquita Canyon landfill in Valencia, and discontinued discharging the city's sludge to the ocean.

Dr. John Dorsey, Hyperion Treatment Plant environmental monitoring manager, was delighted that the city had finally stopped its sludge dumping. He recalled what the area around the sludge outfall was like when he first arrived in 1982. "I called it the 'Black Hole of Calcutta.'" As he explained later, "If you don't get good dispersal and you get a buildup of organic matter in the sea bottom, you get problems." When too much organic matter builds up, then it can't be broken down. "You get hydrogen sulfide and it bonds with iron in the sediment and creates sulfide sediments." So, the area below the 7-Mile Outfall had by 1982 become a seriously degraded habitat. As a result, Dorsey added, "you didn't find very many fish there because their prey items were absent." Prey for the larger fish, such as marine worms, clams, snails, crustaceans and brittle stars could not survive in what had become pretty much of a dead zone. There was life in the sludge field, he said, but only a few species of small marine worms and "a clam with no stomach." That image makes the non-scientist a bit queasy.

In 1991, the city inaugurated a second monitoring vessel to supplement the availability of the *Marine Surveyor*. *La Mer*, the second boat, is a striking eighty-five-foot fiberglass ship that more closely resembles a yacht than the monitoring boat it is. Many of the biologists and chemists who must use *La Mer* to collect organisms, water and equipment samples at approximately sixty sites in Santa Monica Bay prefer to use the *Marine Surveyor*, which they consider a better "working boat." While the *Marine Surveyor* has stabilizers on the side, they are at water level, whereas the stabilizers on *La Mer* are eight feet above the water and create a rock-and-roll effect, causing the most intrepid scientists to experience nausea, even when the seas are calm.

Beginning in June 1992, the city began a series of upgrades to the sewer system. On June 21, 1992, portions of the North Outfall Sewer, extending from La Cienega and Rodeo to the Hyperion plant, were taken offline, so that the seventy-year-old line could be cleaned and repaired, with the installation of reinforced concrete and polyvinylchloride liners to forestall future decay from hydrogen sulfide gas.

The HERS project, which began construction in 1983, was unable to process all the sludge. So, to ensure getting sludge out of the ocean by the December 1989 deadline, the city began a diversified beneficial-use program for the sludge. Contracts for application of the sludge to farmland went to farmers in Yuma, Arizona and Blythe. Ray Kearney, manager of the sludge program, explained that when they started the program, "because of public perception, we were concerned how the public would receive this and we voluntarily restricted all of our farmers to non-food crops, so that no crops that would be consumed by humans could be grown on this soil." In 1990, Kearney's group added San Joaquin Composting to the diversified program. Within two years, Kearney said, "we removed that restriction from our composted products because the marketplace demanded it. The farmers who were buying the compost wanted the material for food crops. So, we talked it over with the environmental groups and they agreed that this would be a safe practice."

By 1993 the city diversified its disposal of thirteen hundred tons of sludge (now called biosolids, in a nationwide effort to apply a friendly public relations face to the heated sludge) a day, with some serving as landfill cover; most of it composted with city tree trimmings at the San Joaquin facility, while thirty-two percent went to farm applications at Yuma and at Blythe. Only eleven percent from the HERS project went to energy recovery. Steam created in the burning process was used for steam turbines that created electric power for the plant. The ash was recovered and used by a copper mine and a cement factory.

A key component of HERS, the Carver-Greenfield Process, kept breaking down. By the end of 1987, multiple problems were reported, including spiral heat exchangers that plugged up and oil that caught fire and created explosions. The heated gas that vented out of the huge tower violated air-quality standards. The five gas turbine engines designed to produce energy from the digester gas never worked properly. The city was forced to use jet engines (originally designed for military helicopters used in Vietnam) because of space limitations on the site. These eventually failed because of the corrosive nature of the gas. Each cost approximately $4.5 million. Later the engines were salvaged for a total of three hundred thousand dollars—the only salvageable elements of the multi-million-dollar HERS project.

On October 2, 1988 a fireball erupted in the hydroextractor and engulfed two workers, who suffered serious burns. The abrasive nature of the sludge continued to plague the system. Finally in 1993, the Carver-Greenfield drying process was abandoned and new, improved steam dryers were started on March 24, 1994.

On December 14, 1996, city engineer Bob Horii pulled the plug on the HERS project, saying that although it was now working, it just cost too much to continue operating. The air pollution systems to control and scrub hydrogen sulfide gas, storage facilities for dewatered sludge and a truck loading facility are all that remain of the ambitious and mammoth HERS project. HERS, for the most part, stands abandoned as an eerie presence at the bustling, highly efficient Hyperion Treatment Plant. Remaining unused is a huge ramp, the equipment for the complex Carver- Greenfield dehydration process, the fluidized bed combusters, the boilers designed to create steam and the steam turbines designed to produce enough energy to run the plant. When the city needs to expand its treatment facilities in the future, these units will probably be demolished.

In 1996, the city began delivering most of its biosolids to a 4,600-acre ranch for land application to non-food crops. Although some would claim that the brown acres the city had created by dumping its sludge into the ocean had just been moved to another location, the city was confident, with federal and state blessings, that this brown sludge was being beneficially used to produce alfalfa and milo, feed for sheep and cattle.

In the late 1980s, the EPA had begun evaluating standards for all use and disposal management of sludge. Dr. Alan Rubin was staff regulator for the EPA for sewage sludge in the United States. As such he oversaw a program that developed comprehensive regulations, issued in 1993, for the use or disposal of sewage sludge. Rubin said that he spent several years consulting with a panel of experts to determine the appropriate application of sludge to the land. In response to concerns about land application, he said, "There has been an incredible amount of anecdotal information that's popped up, primarily in the popular press. This is not necessarily unique to biosolids or sewage sludge. People who live next to garbage dumps, people who live next to microwave towers, people who live next to hog rendering operations, people who live next to factories, incinerators, landfills, et cetera, have always complained about the environment they're living adjacent to." He added, "This is nothing new. The question is, has anybody ever proven, documented any health impacts? And the answer has to be no."

In fact, no studies of consequences of land application, either from living near farms or from ingesting food grown with fertilizer containing biosolids, have been done. The National Biosolids Partnership maintains extensive records of land application and arguments against the practice. By 2007, biosolids application was expanding throughout the country. Most objections come from residents who live near the farms on which the biosolids had been applied. The federal government and almost all states continue to support the practice. Moreover, Rubin notes that there are no limits on metals, dioxin or chemicals in cow manure that is often used as fertilizer. "With sixteen thousand wastewater treatment plants in the United States," Rubin says, "ninety percent are using land application." It's an interesting expansion and replay of sewage use for irrigation and fertilizer in the early decades of the twentieth century.

Peter Machno, a manager with the National Biosolids Partnership, says that the notion of members of national associations of sewage engineers becoming concerned about how sludge "was projected in the media." He explained that the "issue came to a head when Ted Turner announced a new cartoon, *Captain Planet and the Planeteers*, which was a program designed to educate young people about the environment." The cartoon included an evil character named Sly Sludge. Concerned about the characterization, the sewage association wrote to Turner in the hopes of changing the name of the character.

They were unsuccessful with Turner, but more successful with the EPA: when its sludge regulations were issued, "sludge" had become "biosolids." Although this may appear to be a simple propaganda effort, sludge, when heated to either mesophilic (95 degrees Fahrenheit) or thermophilic (131 degrees Fahrenheit) temperature, is changed from its original character, particularly since almost all pathogens and viruses are removed through a twenty-to-thirty-day heating process in digesters at sewage treatment plants. So, all sewage agencies today use the term biosolids to refer to the sludge produced in their digesters.

Los Angeles is currently sending approximately 650 tons of "biosolids" a day to the farm in Kern County. In 2000 the city bought the property and renamed the site *Green Acres*. The farm regularly produces such crops as milo, wheat, corn and alfalfa. Thousands of sheep annually come to feed on the alfalfa crop, and the crops provide feed for dairy farms in the neighborhood as well. The land-applied biosolids are well below federally mandated limits for such heavy metals as arsenic, cadmium, chromium, copper, lead, mercury, molybdenum, nickel, selenium and zinc, with no detectable amounts of PCBs or dioxins, measured at parts per billion, and they meet Class A requirements for pathogens.

American consumers are far more squeamish about using fertilizer made from their own waste than that from animals or crops grown with artificial chemical ingredients. Most food crops are fertilized with either cow and steer manure that frequently contains massive doses of hormones and other feed enhancers.

Kern County voted in June 2006 to oust the city from *Green Acres*. The key concern for Kern County was the water table under the farm. Although city tests showed no contamination, Kern County officials continued to worry that eventually their water supply would be contaminated. In 2007, a state court issued an injunction against the county, but most people expect the county to craft a new ordinance that may stand up to court challenge. If Kern succeeds in kicking the city out, Los Angeles will have to truck the biosolids hundreds of miles away to farms in Arizona and Northern California. The city also has plans for an experimental program to inject sludge deep underground at Terminal Island and recover gas as the sludge degrades. However, results of this project, treating only a small percentage of the sludge, are several years away.

Mayor Bradley retired in 1993 at the age of seventy-six after an unprecedented five terms in office. Although he was initially an opponent of full secondary treatment, by the time he left office the city was committed to improving its sewage treatment. He died on September 29, 1998, two months before the Hyperion Treatment Plant began providing full secondary treatment in November 1998.

In a remarkable turn of events, ten years after the city completed construction of full secondary treatment facilities at Hyperion, the plant continues to meet federal regulations. No other treatment plant in the city's history had ever lasted more than a few years without creating brown acres in the ocean.

In 2008, the plant retains the ability to increase capacity, despite its limited acreage. Although there will always be the occasional, unmistakable musty odor from the plant, mostly due to southwest winds, air emission controls and a variety of odor-scrubbing elements keep the air in Playa del Rey and in El Segundo near the beach clear enough that the surrounding multi-million-dollar homes maintain their value.

To some it may seem strange that after almost a century of pollution, Los Angeles now has garnered every national engineering award possible. Key among them was being named "One of the Top Ten Public Works Projects of the Twentieth Century" by the American Public Works Association. The city shared this award with the Panama Canal, the Bay Area Rapid Transit District (BART), the Grand Coulee Dam & Columbia River Basin Project, the Tennessee Valley Project, the Interstate Highway System, the reversal of the Chicago River in the first decades of the twentieth century, the Hoover Dam-Boulder Canyon, the St. Lawrence Seaway/Power Project and the Golden Gate

Bridge. Not bad for a plant that had polluted Santa Monica Bay for the previous hundred years.

In 2004, Los Angeles settled a lawsuit, initially filed by the Santa Monica Baykeeper in 1998, to stop continued spills from the aging sewer system. Under terms of the settlement agreement, the city was required to "begin work on specific projects to increase the sewer system's capacity," start plans "to assure that the sewer system has sufficient capacity to convey wet-weather flows," initiate "rehabilitation and replacement of the sewer pipes in poor condition," and to submit reports on these plans to include "at least the rehabilitation and replacement of sixty miles of pipe per year on a three-year rolling average and fifty miles of pipe per year."

The city was also required to clean 2,800 miles of pipe on a three-year rolling average, inspect restaurants and improve enforcement of the city's "ordinance regulating the discharge of grease from restaurants, address sewer odors and inspect at least six hundred miles of pipe annually with closed circuit TV." The city was also ordered to pay a fine of eight hundred thousand dollars to be applied to state supplemental environmental projects (SEPs) and to spend an additional $7.7 million on SEPs, bringing the total devoted to SEPs to $8.5 million. The list of projects included wetland and stream restoration programs, located primarily along the Los Angeles River. Finally, the agreement called for "the city to replace 488 miles of sewer lines, clean an additional 2,800 miles every year and institute a program to control pollutants from city streets that end up in the ocean. Los Angeles agreed to allocate two billion dollars to maintain and repair the sewer system over the next ten years."

Two major interceptor projects were completed subsequent to the settlement agreement. The East Central Interceptor Sewer (ECIS) was finished on September 30, 2004 and the Northeast Interceptor Sewer (NEIS) was dedicated on July 21, 2005. ECIS is 11.4 miles of eleven-foot-diameter sewer line winding through the city east of the Los Angeles River, southward along Exposition Boulevard to the North Outfall Sewer connection with the North Outfall Replacement Sewer near the intersection of Jefferson and La Cienega Boulevards. NEIS is a six-mile sewer line that extends northward from the Seventh Street Bridge at the Los Angeles River in Boyle Heights to Eagle Rock Boulevard and San Fernando Road in the Glassell Park area, where it connects to ECIS. These two major projects, each costing several hundred million dollars, have contributed to the reduction of sewage spills in the city. Despite all efforts, spills are an inevitable fact of life in the modern city, but the newest plans, agreed to in the consent decree signed in 2004, will ensure that these spills will be minimized.

Five major sewer lines bring a combined total of approximately 325 MGD to the Hyperion Treatment Plant. They are the Coastal Interceptor Sewer, the North Outfall Relief Sewer, the North Outfall Sewer, the North Central Outfall Sewer and the Central Outfall Sewer. In January 2008, the average sewage service charge for residents was twenty-eight dollars a month.

Afterword

s of this writing, the sewerage system in Los Angeles seems well poised for the inevitable onslaught of new immigrants and restless Americans called to the sun-kissed, temperate climate of the city by the Los Angeles River. The city's sewer system is in the best shape ever. Engineers are repairing and reconstructing all the city's older sewers. All four of the treatment plants can be expanded to meet the needs of the next wave of new residents. The sewers themselves are in the midst of a major rehabilitation program. So, what lies ahead?

Since city engineers no longer have the protection of civil service, politicians will undoubtedly set the direction of sewage treatment in the future. But, if the past is prelude to the future, politicians are almost always behind the curve when it comes to endorsing spending money for the least known and most hidden part of the city's infrastructure. Short-sighted politicians and economic hard times could bring about an unwillingness to expand and improve the city's sewer system.

But another serious issue confronts the city. Southern California is currently experiencing drought. What is more, the population will continue to expand. Water resources are finite. Water conservation can only do so much. Recycling water is not much more expensive than importing water. And federal officials have restricted how much water Los Angeles could import in the past—and they will in the future. City residents produce 450 million gallons of sewage a day. Little more than one percent is recycled. Thirty million gallons per day from Hyperion go to West Basin Municipal Water District, which recycles the water for industrial use and injects it into the South Bay's groundwater basin as a barrier to saltwater intrusion. Some goes to Balboa Lake and a wildlife refuge in the San Fernando Valley, and a very small amount goes to Griffith Park and a nearby mortuary. A minimum of 27 MGD, mandated by the 2004 consent decree, goes to the Los Angeles River, where it provides water at the lower reaches of an extensive wildlife sanctuary. None goes to recharge the water table underneath. This is an outrage. Orange County is recycling 70 MGD of its sewage to eventually make its way into the water supply. Los Angeles County is using 36.83 MGD for groundwater recharge and an additional 37 MGD is used by industry, for landscape and agricultural irrigation, and for wildlife habitat.

The Department of Water and Power determines how much of the city's wastewater will be reclaimed for potable and non-potable reuse. It is a behemoth, whose engineers have always exhibited a hubris that knows no bounds. The first successful water reclamation plant of 1930 was abandoned when the city decided to import water from the Colorado River. After Mayor James Hahn shut down a groundwater program to recycle water from the Tillman Plant because the DWP had not done any public education and people were fearful of "toilet-to-tap," it shut down plans for groundwater recharge of recyclable water. By early 2008, the DWP was planning to conduct a year-long study to find out what residents want to do with recycled water. This, after an extensive survey was completed in 2006 demonstrating overwhelming approval for reuse of recyclable water.

What can be done? Well, Orange County instituted a major education program—not a glitzy advertisement blitz of flyers, brochures, billboards or radio and television, but good old-fashioned, one-on-one discussions with schools, homeowners, community organizations, volunteer associations and the environmental community. And it got widespread approval for recycling wastewater for eventual reintroduction to the water supply. Los Angeles can do the same. The city currently gets its water supply from a combination of groundwater and imported water. At least fifteen percent of the imported water now contains treated sewage discharged from cities in Northern California, Nevada, Colorado and Arizona. It is time to get over the "yuck" factor. We are all drinking some form of recycled water. The newest methods of treating sewage, after primary, secondary and tertiary treatment include microfiltration, reverse osmosis (which pumps water through permeable membranes) and ultraviolet light that removes all contaminants. If bleach is added, the results are purer than distilled water.

Los Angeles has come a long way from its earliest pueblo days when residents thought it would be nifty just to send the sewage to the Los Angeles River. For more than one hundred years, Los Angeles engineers, politicians and the public treated sewage as simply a potential menace, something to be removed—out of sight, out of mind. The ever- growing environmental movement and the powerful presence of such organizations as Heal the Bay and Santa Monica Baykeeper protect the environment. The future of sewage disposal, treatment and reuse in the city will depend on an informed citizenry, willing to accept whatever financial demands may be made to ensure its health and well-being.

Acknowledgments

I am especially grateful to Paddy Calistro and Scott McAuley, my editors at Angel City Press, who gently and persistently guided me through many draft revisions. Working with them has been an exhilarating intellectual experience.

I owe a great debt to retired Bureau of Sanitation Assistant Director James Langley, who tirelessly gave me feedback, read all the chapters in draft and offered the criticism I needed; to Loyola Marine Biology Professor John H. Dorsey, who read the entire manuscript and offered his advice and criticism; and to historian Abraham Hoffman, who read all chapters, offered always-thoughtful comments and made me think more deeply about this project.

In addition, I would like to thank the more than one hundred people who shared their stories about the city's sewers and helped me understand the

◁ Proclaiming "It Took a Legionaire To Do It" and foreshadowing the Doo Dah Parade by several decades, the City Hall Post of the American Legion proudly displays the outfall inspection boat and its one-man crew.

technical aspects of sewage, water reclamation and marine biology. I cannot list them all. I especially want to acknowledge and express my debt to the following people in the Department of Public Works: Tim Haug, Bradley Smith, Wayne Lawson, Chuck Turhollow, Steve Fan, Greg Deets, Homayoun Moghaddam, Harry Sizemore, Veruge Abkian, Chris Smith and Rita Robinson. My gratitude also goes to the Board of Public Works' indefatigable and dedicated Deputy City Attorney Chris Westhoff. Equally generous with their time were Board of Public Works President Cynthia Ruiz, Executive Secretary James Gibson, re-tired City Engineer Robert Horii, retired Bureau of Sanitation Director Delwin Biagi, retired Hyperion Plant Laboratory Manager Joe Nagano, former Board President Maureen Kindel and former Deputy Mayors to Tom Bradley, Mike Gage and Tom Houston. I will always be grateful to the late Mayor Tom Brad-ley, who consented to a lengthy interview in 1986.

Among the dozens of experts who answered my countless questions, thanks are owed to Leon G. Billings, former chief assistant to Senator Edward Muskie, who spoke with me several times about the origins of the Clean Water Act of 1972 and Joseph Moore, program director of the EPA in 1972. Peter Machno with the National Biosolids Partnership and Cecil Lue-Hing, retired director of research and development for the Metropolitan Water Reclamation District of Greater Chicago, provided an understanding of biosolids. I also want to thank Jon Schladweiler of the Arizona Water & Pollution Control Asso-ciation for "Tracking Down the Roots of Our Sanitary Sewers," former EPA Sludge Regulator Alan Rubin, William Van Waggoner with the Department of Water and Power, Orange County Sanitation District's Michael Moore, Los Angeles County Sanitation District's Erle Hartling and Water Replenishment District Program Manager Hoover Ng.

Heal the Bay cofounder Dorothy Green and Felicia Marcus gave gen-erously of their time during the course of more than one interview. I want to thank Santa Monica Baykeeper Executive Director Tracy Egoscue, who gra-ciously shared her experiences with me.

I wish expressly to thank those who, with unflagging patience and con-siderable assistance, have guided me through the thousands of documents con-sulted for this book. I am indebted to the librarians at the Huntington Library, the Seaver Library, the Water Resources Library and the Los Angeles Public Library who assisted me in my search for documents about Los Angeles' sewers. I must also thank, in particular, Dace Taube, Regional History Collection Librarian, Special Collections, at the University of Southern California Doheny Memorial Library; Los Angeles Records Manager Todd Gaydowski; Jay Jones and Mike Holland at the Los Angeles City Archives; Jeffrey Rankin, Simon Elliott, Octavio Olvera and Charles Wilson, cataloguer of the Tom Bradley Administration files at the Department of Special Collections at the University of California, Los Angeles; and El Segundo Deputy City Clerk Cathy Domann and her efficient assistant, Mona Shilling, for indexing and providing access to the historical El Segundo City Council Files.

I am also indebted to Vikki Zale and Chris and Reed Harris, who repeatedly searched their files for photographs and forwarded them to me; Bureau of Sanitation graphics designer Pamela Boddie; Board of Public Works Public Information Officer Lauren Skinner; and to Lynn Thiery, who graciously helped me locate retired Bureau of Sanitation employees. Thanks also to Intellectual Property Counsel Philip H. Lam of the L.A. City Attorney's office. My gratitude to Carolyn See, who encouraged me to keep writing. I also want to thank Robert Gerner and Erika Sloan, whose warmth and integrity have assisted me immeasurably, and William Deverell, who read my original outline and made significant suggestions. I owe a debt of gratitude to Paul Myers for his editing and love of the infrastructure. Thanks also to graphic designer Amy Inouye for turning this overwhelmingly complex story into a visually interesting work.

A very special thanks to Lynn Perdomo, without whose optimism and confidence in me I could not have completed this book. My gratitude for her boundless support goes to Anne Goeke. And, of course, to my children, Jeffrey Sklar and Tracy Sklar, who never stopped believing in this project and who lifted my spirits during some of my more daunting times.

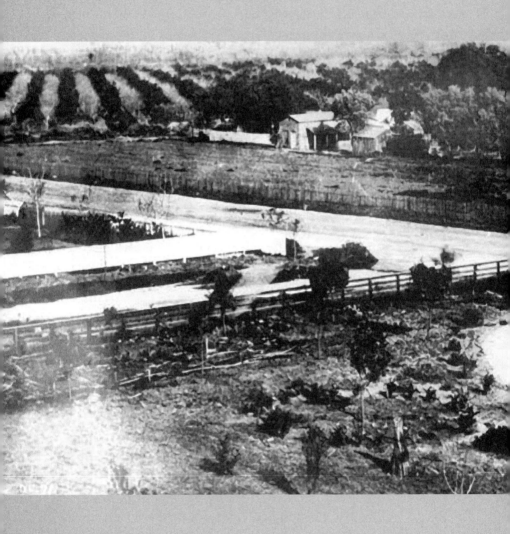

Bibliography

BOOKS

Adamic, Louis. *Laughing in the Jungle: The Autobiography of an Immigrant in America.* New York and London: Harpers & Brothers, 1932.

Asbell, Bernard. *The Senate Nobody Knows.* Garden City, New York: Doubleday, 1978.

Banham, Reyner. *Los Angeles: The Architecture of Four Ecologies.* New York: Harper & Row, 1971.

Bell, Horace. *On the Old West Coast: Being Further Reminiscences of a Ranger.* New York: William Morrow & Co., 1930.

Bowman, Lynn. *Los Angeles: Epic of a City.* Berkeley: Howell-North Books, 1974.

Brook, Harry Ellington. *Irrigation in Southern California.* Los Angeles: Los Angeles Printing, 1893.

Bulloch, David K. *The Wasted Ocean.* New York: Lyons & Burford, 1989.

◁ *Zanjas*, ditches from the Los Angeles River, were long used for both water supply and sewage disposal. circa 1860

Carruth, Gorton, ed. *Encyclopedia of American Facts and Dates.* New York: Crowell, 1972.

Carr, Harry. *Los Angeles: City of Dreams.* New York: D. Appleton-Century, 1935.

Caughey, John and Laree Caughey, eds. *Los Angeles: Biography of a City.* Berkeley: University of California Press, 1976.

Cohen, William A. and Ryan Johnson, ed. *Filth: Dirt, Disgust, and Modern Life.* Minneapolis: University of Minnesota Press, 2005.

Cosgrove, John Joseph. *History of Sanitation.* Pittsburgh: Standard Sanitary Manufacturing, 1909.

Crouch, Winston W., Wendell Maccoby, Margaret G. Morden and Richard Bigger. "Sanitation and Health." In *Metropolitan Los Angeles: A Study in Integration.* Los Angeles: Haynes Foundation, 1952.

Crump, Spencer. *Ride the Big Red Cars: How the Trolleys Helped Build Southern California.* Los Angeles: Crest Publications, 1962.

Davis, Mike. *City of Quartz: Excavating the Future of Los Angeles.* London and New York: Verso, 1990.

———. *Ecology of Fear: Los Angeles and the Imagination of Disaster.* New York: Metropolitan Books, 1998.

DeMarco, Gordon. *A Short History of Los Angeles.* San Francisco: Lexicon, 1988.

Deverell, William. *Whitewashed Adobe: The Rise of Los Angeles and the Remaking of its Mexican Past.* Berkeley and Los Angeles: University of California Press, 2004.

Dumke, Glenn S. *The Boom of the Eighties in Southern California.* San Marino, California: Huntington Library, 1944.

Fogelson, Robert M. *The Fragmented Metropolis, Los Angeles 1850-1930.* Berkeley and Los Angeles: University of California Press, 1967 and 1993.

Folwell, A. Prescott. *Sewerage: the Designing, Constructing and Maintaining of Sewerage Systems and Sewerage Plants.* New York: John Wiley & Sons, 1904.

Goldman, Joanne Abel. *Building New York's Sewers: Developing Mechanisms of Urban Management.* West Lafayette, Indiana: Purdue University Press, 1997.

Gottlieb, Robert and Margaret Fizsimmons. *Thirst for Growth: Water Agencies as Hidden Government in California.* Tucson: University of Arizona Press, 1991.

Graves, Jackson Alpheus. *My Seventy Years in California, 1857-1927.* Los Angeles: Times-Mirror Press, 1927.

Guinn, J.M. *Historical and Biographical Record of Los Angeles and Vicinity: Containing a History of the City from its Earliest Settlement as a Spanish Pueblo to the Closing Year of the Nineteenth Century.* Chicago: Chapman Publishing, 1902.

———. *A History of California*, Vol. 1. Los Angeles: Historic Record Co., 1915.

Hanna, Phil Townsend. *California Through Four Centuries*. New York: Farrar & Rinehart, 1935.

Hendricks, William O. *Moses H. Sherman, A Pioneer Developer and the Pacific Southwest*. Corona del Mar, California: Sherman Foundation, 1971.

Henstell, Bruce. *Sunshine and Wealth: Los Angeles in the Twenties and Thirties*. San Francisco: Chronicle Books, 1984.

Herson, Albert. *Land-based Sewage Sludge Management Alternatives for Los Angeles: Evaluation and Comparison*. Berkeley: School of Architecture & Urban Planning, University of California, 1976.

Hise, Greg. "Los Angeles in the 1920s." In *Metropolis in the Making, Los Angeles in the 1920s*, eds. Tom Sitton and William Deverell. Berkeley and Los Angeles: University of California Press, 2001.

Hoffman, Abraham. *Vision or Villainy: Origins of the Owens Valley-Los Angeles Water Controversy*. College Station, Texas: Texas A & M University Press, 1981.

Hundley, Norris, Jr. *The Great Thirst: Californians and Water, 1770s-1990s*. Los Angeles and Berkeley: University of California Press, 1992.

Hunter, Burton L. *Evolution of Municipal Organization and Administrative Practice in the City of Los Angeles*. Los Angeles: Parker Stone, & Baird, 1933.

Huxley, Aldous. "Hyperion to a Satyr." In *Tomorrow and Tomorrow and Tomorrow*. New York: Harper & Row, 1956. 149-71.

Ingersoll, Luther A. *Ingersoll's Century History, Santa Monica Bay Cities: Prefaced with a Brief History of the State of California, a Condensed History of Los Angeles County, 1542-1908; Supplemented with an Encyclopedia of Local Biography*. Los Angeles: L.A. Ingersoll, 1908.

Jackson, George, Robert C.Y. Koh, Norman H. Brooks, James J. Morgan. *Assessment of Alternative Strategies for Sludge Disposal into Deep Ocean Basins off Southern California*. Pasadena: Environment Quality Laboratory, California Institute of Technology, 1979.

James, George Wharton. *Tourists' Guide to Southern California*. Los Angeles: Chamber of Commerce, 1894.

Johnson, Steven. *The Ghost Map: The Story of London's Most Terrifying Epidemic—and How It Changed Science, Cities, and the Modern World*. New York: Penguin, 2007.

Kahrl, William L. *Water and Power: The Conflict Over Los Angeles' Water Supply in the Owen's Valley*. Berkeley and Los Angeles: University of California Press, 1982.

Klein, Norman M. *The History of Forgetting: Los Angeles and the Erasure of Memory*. London and New York: Verso, 1998.

Lindley, Walter, M.D. and J. P. Widney, A.M., M.D. *California of the South: Being a Complete Guide-book to Southern California.* New York: D. Appleton and Company, 1888.

Ling, Susie, ed. *Bridging the Centuries: History of Chinese Americans in Southern California.* Los Angeles: Chinese Historical Society of Southern California, 2001.

Lue-Hing, Cecil, David R. Zenz and Richard Kuchenrither, eds. *Municipal Sewage Sludge Management: Processing, Utilization, and Disposal.* Lancaster, Pennsylvania: Technomic Pub., 1992.

Marshall, Robert. *In the Sewers of Lvov: A Heroic Story of Survival From the Holocaust.* New York: Scribner's, 1991.

McGroarty, John Steven. *History of Los Angeles County.* Chicago: American Historical Society, 1923.

McCuen, Gary E., ed. *Protecting Water Quality.* Hudson, Wisconsin: G.E.M. Publications, 1986.

McWilliams, Carey. *California: The Great Exception.* New York: Current Books, Inc., 1949.

———. *Southern California Country: An Island on the Land.* New York: Duell, Sloan & Pearce, 1946.

Mayer, Robert. *Los Angeles: A Chronological and Documentary History 1542-1976.* Dobbs Ferry, New York: Oceana Publications, 1978.

Melosi, Martin V. , ed. *Pollution and Reform in American Cities, 1870-1930.* Austin: University of Texas Press, 1980.

———. *The Sanitary City: Urban Infrastructure in America from Colonial Times to the Present.* Baltimore: Johns Hopkins University Press, 2000.

Metcalf, Leonard and Harrison P. Eddy "The Lessons Taught by Early Sewerage Works." In *American Sewerage Practice,* rev. by Harrison P. Eddy. 3rd edition. New York: McGraw-Hill, 1935.

———. *Sewerage and Sewage Disposal: A Textbook.* New York: McGraw-Hill, 1922.

Morrison, Patt. *Río L.A.: Tales From the Los Angeles River.* Santa Monica: Angel City Press, 2001.

Mulholland, Catherine. *William Mulholland and the Rise of Los Angeles.* Los Angeles and Berkeley: University of California Press, 2000.

Myers, William A. and Ira L. Swett. *Trolleys to the Surf: The Story of the Los Angeles Pacific Railway 63.* Glendale, California: Interurbans, 1976.

Nadeau, Remi. *California, the New Society.* New York: David McKay, 1963.

———. *City Makers: The Story of Southern California's First Boom, 1868-1876*. Corona del Mar, California: Trans-Anglo Books, 1965.

———. *Los Angeles: From Mission to Modern City*. New York: Longmans, Green, 1960.

———. *The Water Seekers*. Garden City, New York: Doubleday, 1950.

Nelson, Howard. *Los Angeles Metropolis*. Dubuque, Iowa: Kendall/Hunt, 1983.

Newmark, Maurice and Marco Newmark, eds. *Sixty Years in Southern California: Containing the Reminiscences of Harris Newmark, 1853-1913*. New York: Knickerbocker Press, 1916.

O'Flaherty, Joseph S. *Those Powerful Years: The South Coast and Los Angeles, 1887-1917*. Hicksville, New York: Exposition Press, 1978.

Orsi, Richard J. *Sunset Limited: The Southern Pacific Railroad and the Development of the American West 1850- 1930*. Berkeley and Los Angeles: University of California Press, 2005.

Outwater, Alice. *Water: A Natural History*. New York: Basic Books, 1996.

Ovnick, Merry. *Los Angeles: The End of the Rainbow*. Los Angeles: Balcony Press, 1994.

Patrick, Ruth. *Surface Water Quality: Have the Laws Been Successful?*. Princeton: Princeton University Press, 1992.

Petulla, Joseph M. *Environmental Protection in the United States, Industry, Agencies, Environmentalists*. San Francisco: San Francisco Study Center, 1987.

Pitt, Leonard and Dale Pitt. *Los Angeles A to Z: An Encyclopedia of the City and County*. Berkeley and Los Angeles: University of California Press, 1997.

Rawn, A.M. *Narrative—C.S.D.* Los Angeles: County Sanitation Districts of Los Angeles County, 1965.

Reid, Donald. *Paris Sewers and Sewermen: Realities and Representations*. Cambridge: Harvard University Press, 1991.

Reynolds, Terry S., ed. *The Engineer in America*. Chicago: University of Chicago Press, 1991.

Reisner, Marc. *Cadillac Desert: The American West and Its Disappearing Water*. New York: Viking, 1986.

Robinson, W. W. *Inglewood: A Calendar of Events in the Making of a City*. Los Angeles: Title Insurance and Trust, 1947.

Royte, Elizabeth. *Garbage Land: On the Secret Trail of Trash*. New York: Little Brown, 2005.

Rudd, Hynda, ed. *Los Angeles and Its Environs in the Twentieth Century: a Bibliography of a Metropolis, 1970-1990, with a Directory of its Resources in Los Angeles County*. Los Angeles: Los Angeles Historical Society, 1996.

Sitton, Tom. *Los Angeles Transformed: Fletcher Bowron's Urban Reform Revival, 1938-1953*. Albuquerque: University of New Mexico Press, 2005.

Spalding, William Andrew. *History and Reminiscences: Los Angeles City and County, California*. Los Angeles: J.R. Finnell & Sons, 1931.

Starr, Kevin. *Inventing the Dream: California Through the Progressive Era*. New York: Oxford University Press, 1985.

———. *Material Dreams: Southern California through the 1920s*. New York: Oxford University Press, 1990.

Swett, Ira L. *Los Angeles Pacific Album 18*. Glendale, California: Interurbans, 1955.

———. *Los Angeles Pacific Album 40*. Glendale, California: Interurbans, 1965.

Tarr, Joel A. *The Search for the Ultimate Sink: Urban Pollution in Historical Perspective*. Akron, Ohio: University of Akron Press, 1996.

——— and Gabriel DePuy, eds. *Technology and the Rise of the Networked City in Europe and America*. Philadelphia: Temple University Press, 1988.

Widney, J.P. *The Lure and the Land: an Idyl of the Pacific*. Los Angeles: Pacific Publishing, 1932.

Wiley, Peter and Robert Gottlieb. *Empires in the Sun: The Rise of the New American West*. Tucson: University of Arizona Press, 1982.

Vance, Mary. *Sewers and Sewerage: A Bibliography*. Monticello, Illinois: Vance Bibliographies, 1982.

Walton, John. *Western Times and Water Wars: State, Culture and Rebellion in Southern California*. Berkeley and Los Angeles: University of California Press, 1992.

White, Leslie T. *Me, Detective*. New York: Harcourt, Brace, 1936.

Wenner, Lettie M. *The Environmental Decade in Court*. Bloomington: Indiana University Press, 1982.

ARTICLES AND REPORTS

Belt, Elmer, M.D. "Sanitary Survey of Sewage Pollution of Santa Monica Bay," *Western City* (June 1943): 17-22.

Boroughs, Reuben. "Sewage: Shame of a City," *Civic Affairs* (1945).

Burian, Steven J., Stephan J. Nix, Robert E. Pitt, and S. Rocky Durrans. "Urban Wastewater Management in the United States." *Journal of Urban Technology* 7, No. 3, 33-62.

Engineering-Science, Inc. *Offsite Sludge Transportation and Disposal Program*, Draft EIR, (August 1988).

Los Angeles Chamber of Commerce. *Sixty Achieving Years, 1888-1948*. University of California, Los Angeles Special Collections.

——. *Facts and Figures Concerning Southern California and Los Angeles City and County*, (December 1888).

Knowlton, Willis T. "Los Angeles Sewage Disposal Plans Assuming Final Form," *Engineering News Record* 90 (June 7, 1923).

——. "The Sewage Disposal Problem of Los Angeles, California," *Transactions of the American Society of Civil Engineers* 92 (1928): 984-993.

Parkes, G.A. "Los Angeles New 245 M.G.D. Hyperion Treatment Plant Nears Completion." *Western City* (April 1950).

Peterson, Jon A. "The Impact of Sanitary Reform Upon American Urban Planning, 1840-1890." *Journal of Social History* 13 (Fall 1979): 83-103.

Rittenberg, Sidney. "Studies on Coliform Bacteria Discharged From the Hyperion Outfall Final Bacteriological Report." A final report submitted by the University of Southern California (August 29, 1956).

Santa Monica Bay Restoration Project. *State of the Bay; Actions for Bay Restoration—Draft*; miscellaneous reports and papers, 1988-1994. (April 1994).

Smith, H.G. "Proposed Changes and Improvements in Sewage Disposal at the Hyperion Plant of Los Angeles City." *California Sewage Works Journal* 9 (1937): 33-42.

Tarr, Joel, James McCurley, and Terry F. Yosie. "The Development and Impact of Urban Wastewater Technology: Changing Concepts of Water Quality Control, 1850-1930." *Pollution and Reform in American Cities, 1870-1930*, ed. Martin V. Melosi. University of Texas Press, (1980): 59-82.

GOVERNMENT DOCUMENTS AND REPORTS

California, State of. Report of the State Engineer to the Board of Directors of the Stockton Insane Asylum. *The Sewage Question in California*. By William Hamilton Hall. 1883.

——. California State Board of Public Health. *Report on A Pollution Survey of Santa Monica Bay Beaches in 1942, Bureau of Sanitary Engineering*. June 26, 1943.

———. Regional Water Pollution Control Board, No. 4 (Later renamed Los Angeles Regional Water Quality Control Board)— *Reports and Miscellaneous Files and Records,* 1952-1994.

———. *The Citizens Forum on Sludge,* September 1977.

LA/OMA Project Staff. *Sludge Processing and Disposal.* Los Angeles, April 1977.

Los Angeles, City of. *3. Compiled Ordinances and Resolutions of the City of Los Angeles.* Compiled and indexed by Freeman G. Teed. 1887.

———. *4. Charter and Compiled Ordinances and Resolutions of the City of Los Angeles.* Compiled and indexed by Freeman G. Teed. 1889.

———. Board of Consulting Engineers. *Report Upon a Program of Sewerage and Sewage Treatment and Disposal for the City of Los Angeles, California and Certain of Its Environs.* By Samuel A. Greeley, Charles G. Hyde and Franklin Thomas. November 1939.

———. Board of Consulting Engineers. *Review of and Report on Plans for an Ocean Outfall for the City of Los Angeles at Hyperion.* By Charles T. Leeds, A.M. Rawn, Franklin Thomas. 1974.

———. Board of Engineers. *Report to City Council.* By Rudolf Hering, George C. Knox, August Mayer, Fred Eaton. University of California Special Collections. 1889.

———. Board of Engineers. *Review of Plans for an Ocean Outfall for the City of Los Angeles at Hyperion.* By Leeds, Rawn, Thomas. 1946.

———. Board of Public Service Commissioners, Department of Public Services. *Aqueduct Final Report.* 1916.

———. Board of Public Works. *Annual Reports.* 1956-1991.

———. Bureau of Engineering, Department of Public Works. *Annual Reports.* 1900-2005.

———. Bureau of Engineering. *Proposed Six-Year Program of Post-War Construction for the City of Los Angeles.* By Lloyd Aldrich. 1944.

———. Bureau of Engineering. *Sludge Disposal Alternatives for the City of Los Angeles.* By Donald C. Tillman, City Engineer. May 1973.

———. Bureau of Engineering. *Survey of Unemployment Relief Activities in the City of Los Angeles.* 1938. ———. Bureau of Engineering, Department of Public Works. *Wastewater Facilities Plan.* November 1977.

———. Bureau of Sanitation, Department of Public Works. *Annual and Miscellaneous Reports.* 1951-2005.

———. *City Engineer Reports.* 1870-1900.

———. Clean Water Program, Department of Public Works. *Advance Planning Report.* 1993.

——. *Common Council Records. Petitions, Minutes, City Council Minutes*, Los Angeles City Archives. 1850-2005.

——. Department of Public Works. *City Engineers, 1855-1981*, ed. John P. Hunt and Bernice Kimball. City Engineer. 1981.

——. Department of Public Works. *Report on Waste Discharges to the Ocean.* 1974.

——. Department of Water & Power. *Wastewater Reclamation by the City of Los Angeles.* Oct. 10, 1968.

——. *Los Angeles City Officials*, City Library, Municipal Reference Department. 1850-1965.

——. *Municipal Records.* 1879-1886.

——. Sewage Disposal Committee. *Report of Engineers Regarding the Disposal of Sewage of the City of Los Angeles California.* October 1921.

——. *Technical Report to Board of Public Works Upon the Sewage Disposal Problem of Los Angeles and Associated Communities.* By Metcalf & Eddy, engineers. 1944

.——.Wastewater Systems Engineering Division, Bureau of Engineering, Department of Public Works. *Project Report Hyperion Treatment Plant Sludge Processing and Disposal System.* March 1974.

——. Wastewater Systems Engineering Division, Bureau of Engineering, Department of Public Works. *Project Report Hyperion Treatment Plant Interim Sludge Processing and Disposal System.* April 1975.

Los Angeles, County of. *Report upon the Potential Reclamation of Sewage Now Wasting to Los Angeles County.* By H.E. Hedger, A.M. Rawn. 1958.

——. *Report upon the Reclamation of Water from Sewage and Industrial Waste in Los Angeles County.* By C.E. Arnold, H.E. Hedger, A.M. Rawn. 1949.

United States Bureau of the Census. *General Statistics of Cities: 1909.* Washington, 1913. 18-35.

——. 1. Fifteenth Census of the United States: *Population.* Washington, 1931. 18-19.

——. 95th Congress, 2nd Session, Serial No. 95-14. *A Legislative History of the Clean Water Act of 1977.*

——. Office of the Comptroller General. Report to Congress. *Problems and Progress in Regulating Ocean Dumping of Sewage Sludge and Industrial Wastes.* January 21, 1977.

——. U.S. Government Printing Office. Tenth Census of the United States, 1880, *Report on the Social Statistics of Cities.* By George E. Waring, Jr. 1886.

Southern California Association of Governments. *Area-wide Waste Treatment Management Plan.* Los Angeles: SCAG. April 1979.

PAMPHLETS AND UNPUBLISHED MANUSCRIPTS

Blake, Aldrich and Ernest R. Chamberlain. *Sewers of Los Angeles: Millions in Damages . Disease . Pestilence . Death.* Ernest R. Chamberlain Collection, University of California, Los Angeles, Special Collections Library. 1943.

Borough, Reuben W. *Papers 1900-1970.* University of California, Los Angeles, Special Collections Library.

———. Oral History Program, University of California, Los Angeles, Special Collections Library.

Chamberlain, Ernest R. *Papers.* University of California, Los Angeles, Special Collections Library.

Clark, Margarete. Oral History Program. University of California, Los Angeles, Special Collections Library.

Freeman, Daniel. *Papers.* University of California, Los Angeles, Special Collections Library,

Kitahara, Michael R. and Mattas, Steven T. *Enforcement Under the Clean Water Act: A California Waste Discharge Survey.* M.A. Thesis, UCLA School of Architecture and Urban Planning. 1988.

Layne, J. Gregg. Los Angeles Department of Water and Power. *Water and Power for a Great City.* unpublished manuscript. 1957.

Lippincott, Joseph Barlow. *Operation of the Los Angeles Out-Fall Sewer and Sewage Irrigation.* University of California Berkeley Water Resources Center Archives. 1897.

Nikolitch, Milan. *Sewerage System of Los Angeles and Sewage Irrigation in Southern California.* M.S. Thesis, University of California Berkeley. 1906.

Poulson, Norris. Oral History Program. University of California, Los Angeles, Special Collections Library.1965.

Sherman, Moses, H. *Papers.* Sherman Foundation Center, Corona Del Mar, California.

Tenney, Jack B. Oral History Transcript. Special Collections Library, University of California, Los Angeles. 1969.

Van Norman, James Howe. Oral History Program. Special Collections Library, University of California, Los Angeles. 1989.

Index

Activated sludge treatment, 62, 106, 124
Aeration tanks, 109, 116, 158, 159
Agricultural Park, 26
Aldrich, Lloyd, 86, 96, 105, 113, 114, 121
Alkyl benzene sulfonate (ABS), 115, 125
Allan Hancock Foundation, 118
American Railway Workers Union, 49
American Society of Civil Engineers, 116
Applied Recovery Technology, 200
Atkinson, Guy, 111
Balantidium coli (B.coli), 76
Ballona Creek, 15, 26, 30, 64, 68-70, 75, 95, 97, 99, 101, 111, 118, 123, 132, 149, 173, 186, 187, 190
Ballona Harbor Company, 31
Bascom, Willard, 180-181, 185-186
Belgrand, Eugene, 15
Belleview Outfall, 57, 64, 71
Bell, Horace, 20
Belt, Elmer, 102

Bennett, Howard, 150, 179, 183
Betz, Jack, 148
Biagi, Delwin A., 187, 192-193, 197
Bielenson, Anthony, 180, 185
Biosolids, 176, 203-206
Blake, Aldrich, 104
Board of Public Works, 50-51, 53, 56, 58, 60, 68-71, 74-75, 93, 95, 100-102, 106-108, 110, 112, 118, 121, 127-128, 131, 142, 146-147, 181, 189, 191-192, 199-201
Borough, Reuben, 61, 63-64, 101, 110
Bowron, Fletcher, 99, 110, 114, 129, 151, 157
Bradley, Tom, 141, 181
Brown, David, 186
Brown, Jennie, 50, 52
Brown, Kathleen, 199
Brown, Reuben, 98
Bureau of Engineering, 94, 101, 112-113, 127, 133, 148-149, 187, 191-192

Bureau of Sanitation, 101, 112-113, 127, 131, 146, 149, 187, 191-192, 195, 200
Bureau of Street Maintenance, 101
Bureau of Water Works and Supply, 61, 70
C. Forrester Company, 46, 47
California Institute of Technology, 101, 107
California State Board of Health, 42, 53, 55-57, 61, 74, 76, 94, 101-106, 111-114, 116, 124
Carson, Rachel, 125
Carver Greenfield Process, 145-146
Centinela Ranch, 26, 30, 34, 38
Central Outfall Sewer (COS), 6, 12, 43, 57, 69, 75, 132, 152-153, 172, 208
Chamberlain, Ernest, 104
Chandler, Harry, 49, 58
Chappell, Howard, 131
Chemfix, 201
City Council Sewer Committee, 50
Civil service, 50, 71, 99, 101, 209
Civil Works Administration, 96
Clark, Eli, 48
Clean Water Act of 1972, 139-140, 146, 196
Cleveland, Grover, 49
Clinton, Clifford, 99, 104
Cloaca Maxima, 14
Cloacina, 14
Coalition to Stop Dumping Sewage into the Ocean, 184
Coastal Interceptor Sewer (CIS), 6, 208
Colorado River, 75, 95, 210
Crouch, C.D., 62
Cryer, George, 93
Culver City News, 185
Daily Mirror, 111
Daily News, 111
Daniel Freeman, 30, 39-40, 42, 48
Daniel Mann Johnson and Mendenhall (DMJM), 122
DeFalco, Paul, 141, 146
Department of Engineering, 36, 59, 69, 99, 101
Department of Water and Power, 70, 94, 111, 195, 210
Derby, C.F., 39
Desert Bloom, 141
Deukmejian, Gov. George, 190, 192

Dichloro diphenyl trichloroethane (DDT), 125, 126, 142, 181, 185
Dickey Water Pollution Act, 113
Dockweiler Beach, 26, 114
Dockweiler, Frederick, 114
Dockweiler, Henry, 37, 114
Dockweiler, Isadore, 114
Dockweiler, John Henry, 37, 114
Dockweiler Sewer, 39, 42-43, 48, 59
Donald C. Tillman Water Reclamation Plant, 16, 147, 191
Dorsey, John, 201
East Central Interceptor Sewer (ECIS), 177, 207
Eaton, Fred, 24-25, 28, 32-35, 37, 68, 79, 96, 98
Eddy, Harrison P., 22
Edwards Air Force Base, 141
Edwards, S.J., 47
Engineering News Record, 60
Environmental Protection Agency, 134, 139
Escherichia coli (E. coli), 76
Express, The, 52
Fabiani, Mark, 189
Fay, Rimmon, 149, 178, 179
Ferraro, John, 144
Ferrell, David, 183
Ferris, Horace B., 128
Film With an Odor, The, 64
5-Mile Outfall, 6, 118, 119, 122, 124-125, 167, 192
Ford, John Anson, 99
Fuller, George W., 61
Gage, Mike, 197, 199
Garber, William, 146
Gault, Charlayne Hunter, 139
Gillelen, Frank, 106
Gill, Louis Dodge, 131, 133
Gladstein, Cliff, 181
Good Government movement, 50, 59
Goudey, Ray, 94
Graham, E.H. , Jr., 122
Greeley, Samuel, 101
Green, Dorothy, 172, 179, 184, 195, 197, 232
Griffin, John Alden, 68, 72
Griffith Park, 94-95, 116, 117, 210

Grossman, James, 184
Groveman, Barry, 200
Hamlin, Homer, 52, 59
Hansen, Andrew, 59
Haug, Roger T. (Tim), 146
Hayden, Tom, 181, 185, 187
Haynes, John Randolph, 5, 10, 50
Heal the Bay, 4, 172, 178, 190, 194, 211, 232
Heavy metals, 140, 183, 205
Hering, Rudolf, 28, 32, 41
Hing, Cecil Lue, 16
Holmes & Narver, 122
Honeymooners, The, 9
Horii, Robert, 181, 184, 196, 199, 203
Houston, Tom, 197, 200
Hume, Norm, 131
Huxley, Aldous, 114
Hyde, Charles, 101
Hyde, William, 93
Hyperion, 2, 16, 48-50, 55-60, 62, 71-73, 75-
   78, 85, 89, 94-99, 101-107, 110-120,
   122-127, 133, 136, 142, 145-146, 148-
   149, 154, 157-161, 166, 172, 175, 182,
   184, 188, 191-196, 201-203, 206, 208,
   210
Hyperion Energy Recovery System (HERS),
   50, 145-146, 148-149, 188-189, 191,
   193-194, 196, 200, 202-203
Hyperion Engineers, 122
Hyperion Screening Plant, 2, 85, 104, 115, 123
Hyperion Treatment Plant, 16, 110, 112, 123-
   124, 136, 154, 159-160, 172, 182, 201,
   203, 206, 208
I Love Lucy, 113
Independent Brick Company, 46, 47
Interceptors, 6, 16, 26, 28-29, 32, 126, 132
Jessup, John, 93
John Randolph Haynes and Dora Haynes
   Foundation, 5, 10
Johnson, Hiram, 129
Kennedy, Clyde C., 118
Kindel, Maureen, 181, 184, 187-190, 192, 195,
   197, 199
Kinney, Abbott, 31
Knowlton, Willis T., 56, 68
Knox, George, 23

Koebig Sons, 122
Kuchel, Thomas, 129
Kurtz, Carol, 181
La Cienega Rancho, 53
La Cienega San Fernando Valley Relief Sewer
   (LCSFVRS), 6, 74, 126
Lambie, William, 29
La Mer, 170, 202
Laterals, 16, 26
Launders 109
Leask, Samuel, 127, 128
Lecouvreur, Robert L., 20
Leeds and Hill, 108
Levine, Mel, 185
Linear alkylate sulfonate (LAS), 125
Lippincott, J.B., 40
Los Angeles/Orange County Metropolitan Area
   (LA/OMA), 142, 145
Los Angeles Aqueduct, 25, 35, 56, 68, 70
Los Angeles City Water Company, 25, 38
Los Angeles County Farm Bureau, 64
Los Angeles Glendale Water Reclamation
   Plant, 16, 147, 169
Los Angeles Herald, 31, 185
Los Angeles Independent, 111
Los Angeles Pressed Brick Company, 47
Los Angeles Public Works Department, 10
Los Angeles River, 15, 17, 19-21, 32, 34-35,
   49, 60, 62, 68, 75, 95, 101, 106, 113,
   116-117, 120, 147, 165, 169, 191, 207,
   209-211, 216
Los Angeles Times, 29, 31, 39, 49, 52, 62, 70,
   74, 96, 101, 103, 110-111, 114, 143, 147,
   179, 183, 185, 188, 232
Louis, Nowell, 134, 143, 144
Ludwig, Kenneth, 192
Lumsden, Anthony, 147
Machno, Peter, 205
Marcus, Felicia, 180, 184, 189, 194, 197, 199
Marine Surveyor, The, 170, 202
May, Don, 180
Mayer, August, 32
McAleer, Owen, 50
McFarland, Dan, 26
McWilliams, Carey, 67
Merchants and Manufacturers Association, 49

Metcalf & Eddy, 106, 107, 117
Metcalf, Leonard, 22
Metropolitan Sanitation District, 73
Metropolitan Water District, 95
Milorganite, 63
Montrose Chemical Corporation, 126
Mulholland, William, 25, 34, 50, 51, 68, 70
Municipal League Bulletin, 61
Muskie, Sen. Edmund S., 139
Nadeau's Vineyard, 32
Nagano, Joe, 120, 232
Narver, David L., 122
National Biosolids Partnership, 204, 205
National Environmental Policy Act, 139
National Pollutant Discharge Elimination
    System (NPDES), 140, 143, 144, 194
Newmark, Harris, 19
Nishikawa, Dennis, 200
Nixon, Richard M., 134
North Central Outfall Sewer (NCOS), 6, 99,
    101, 103, 119, 122-124, 132, 208
Northeast Interceptor Sewer (NEIS), 207
North Outfall Replacement Sewer (NORS),
    6, 207
North Outfall Sewer (NOS), 6, 56, 60, 66, 71,
    74, 75, 81, 88, 94, 96, 100, 149, 202,
    207, 208
North Outfall Treatment Facility, 190
O'Hara, George, 146, 148, 193
Ofner, Milton, 128
Old Faithful, 63, 64
Olmstead, Frank, 41
1-Mile Outfall, 6, 98, 112, 116, 123-125, 148-
    149, 163, 187, 192
Otis, Harrison Gray, 49
Owens Valley, 34, 35, 57, 68
Pacific Electric Railway, 48
Pacific Legal Foundation, 144
Pacific Sewerage Company, 32
Pardee, Lyle, 121-122, 131-132, 141
Parker, John, 138
Piper, C. Irwin, 131
Pollack, Mark, 180
Polychlorinated biphenyls (PCBs), 142
Pomeroy, Hugh, 73
Pomeroy, Richard, 118

Porter Cologne Act, 138-140
Porter, John C., 93
Poulson, Norris, 114, 129
Pregerson, Harry, 180, 181, 188-189, 195
Primary treatment, 108, 111, 119, 124, 142
Progressives, 50, 59
Prowler, The, 119, 170
Public Works Administration, 96
Rawn, A.M., 73, 108, 112, 117
Regional Water Pollution Control Board, 113,
    117
Roosevelt, Franklin D., 96
Rosenthal, Herschel, 185
Rubin, Alan, 204
Russell, Pat, 181, 184
San Fernando Valley, 6, 17, 56, 58, 60, 95, 99-
    101, 113-114, 116, 126-127, 131, 133,
    141, 149, 191, 210
San Gabriel River, 49
Santa Barbara Channel, 137
Santa Fe Railway, 23
Santa Monica Bay, 6, 8, 10, 15, 48, 75, 87, 93,
    102-103, 118, 125, 126, 139, 150, 166-
    167, 184-185, 188, 196, 202, 207
Santa Monica Board of Trade, 31
Santa Monica Daily Outlook, 50
Santa Monica Outlook, 185
Savage, Fran, 201
Schladweiler, Jon C., 14
Schneider, Warren, 101, 112
Schwendinger, Royal O., 146
Secondary treatment, 112, 119, 122, 124, 134,
    139, 142, 144, 146, 149, 180-181, 184-
    186, 189-190, 192, 194, 196, 206
Sepulveda Water Reclamation Plant, 134
service laterals, 16
7-Mile Outfall, 6, 118, 122-124, 141, 166-167,
    183, 202
Shaw, Frank, 96, 105, 114, 129
Sherman, Moses, 48
Simons Brick Company, 44, 46
Simons, Elmer, 46
Simons, Jamie, 180
Simons, Reuben, 46
Simons, Walter, 46
Sizemore, Harry, 188, 193, 200

Sludge, 22, 61, 63, 77, 94-95, 97, 106, 108-109, 114-118, 122-124, 139-146, 148-149, 156, 166-167, 176, 183, 186, 188-189, 194, 196, 200-206
Sludge digesters, 145, 156
Smith, H.G., 73
Smith, Homer, 75
Snow, Dr. John, 15
Snyder, Art, 143
Southern California Coastal Water Research Project (SCCWRP), 138, 142, 180, 185, 191
Southern Pacific Railway, 21
South Side Irrigation Company, 23, 33
Spence, Edward, 31
Stafford, Harry F., 42
Stansbury & Powell, 47, 49, 51-52
State Water Resources Control Board, 138-140, 143
Stavnezer, Moe, 180
Submarine outfall, 63, 72, 76, 94, 98-99, 106-107, 110-111
Suicide, 50, 52
Terminal Island, 16, 74, 87, 96, 126, 174, 193, 206
Terminal Island Treatment Plant, 16, 96, 126, 174
Tertiary treatment, 211
Thomas, Franklin, 101, 107, 112
Tillman, Donald C., 16, 126, 127, 141-142, 146-147, 191
Travel Town, 117
U.S. Army Corps of Engineers, 133, 147
Vail, Hugh, 26
Valenzuela, Ralph, 200
Valley Settling Basin, 117, 147
Van Norman, Harvey, 70, 110
Van Norman, James, 76-78
Vidaure, Oscar Padilla, 200
Vinicio Cerezo Arévalo, Marco, 200
Wada, Frank, 113, 117
Walter Lindley, 29
Waring, George E., 25
Wave, The, 185
Webber, Ernest, 131
West Basin Municipal Water District, 210

Westhoff, Chris, 200
Whipple, George C., 61
White Point, 74, 108, 117, 126, 185
White Point Outfall, 126
Widney, Joseph, 29
Wood, J. Perry, 69
Workman, William, 24
Works Progress Administration, 91, 96-97
Yorty, Sam, 129, 134, 141
Zanjas, 19, 22, 216

Photo Credits

Except where noted below, all photographs in this book are from the collections of City of Los Angeles Archives and the City of Los Angeles Department of Public Works, and are reprinted with the kind permission of the City of Los Angeles.

Thanks also to:

Heal the Bay: 172 (Dorothy Green)

The Huntington Library, Art Collections, and Botanical Gardens: 44

Los Angeles Public Library: 165

Los Angeles Times: 178

Ernest Marquez Collection: 212

Private collection of Joe Nagano: 85 (tile), 136, 154-155, 168

Natural History Museum of Los Angeles County, Seaver Center for Western History Research: 85 (South Screening Plant)

Private collection of Anna Sklar: 151, 162, 163, 166, 167, 170, 171 (*Marine Surveyor*)

University of Southern California, Special Collections Library, California Historical Society Collection: 90 (inspector), 159

Banners above downtown exhort voters to address the sewage problems underfoot.